Praise for *He*

'A glorious anthem to the hedge! The sleeping giant that is waiting to be nurtured back to life-giving health. This wonderful book fires the imagination and will change the way you look at a hedge forever. And never was there a more urgent time to restore our degraded hedges into the lifelines they should be – for our beleaguered wildlife and for ourselves. Endlessly fascinating and teeming with 'Wow!'s. A must-read for anyone who wants to help restore ecological health to our countryside.'

Keggie Carew, author of *Beastly*

'A hugely enjoyable ramble down the tangled green lanes of the beleaguered British countryside. Christopher Hart's investigations shine a new light on the humble hedgerow, in which all manner of solutions to our environmental problems can be found.'

Lee Schofield, author of *Wild Fell*

'This is a soaring love song to the hedge – not the straight-shorn sanitised garden hedges but the ancient hedgerows that heave with life. Hart is an enthusiastic guide to these beautiful, underestimated ecological niches, which not only provide habitats for birds, invertebrates and plants but are also carbon banks and food stores, and even supply ingredients for cocktails.'

Andrea Wulf, author of *Magnificent Rebels*

'We now know that if laid end to end, English hedges would stretch ten times around the earth. In *Hedgelands*, Christopher Hart has put his arms around the whole extraordinary network. He inspires us to see that the hedgerow is a national treasure even more important than we realised.'

Mark Cocker, author and naturalist

'*Hedgelands* is brilliant – fascinating, inspiring, important and extremely timely. It has taken my love and appreciation of our precious hedges to another level. Please buy it, read it and share it with everyone you know who has enough space to plant a hedge!'

Brigit Strawbridge Howard, author of *Dancing with Bees*

'A wonderful history of the threads that stitch the British countryside together, *Hedgelands* provides a detailed description of the wonder of hedges and the vast array of species and benefits they provide for us.'

Jake Fiennes, author of *Land Healer*

'A lovely book on a really important subject. Passionate, personal and important. You won't ever look at a hedgerow in the same way again.'

Roger Morgan-Grenville, author of *The Return of the Grey Partridge*

'This deeply enjoyable book makes me want to go live in a mighty conservation hedge – or at least create one. Hart has delved deep into fascinating hedge etymology and history, and brings the joy of restoring a hedge to life. As an insect lover, I welcomed the stories of renewal and abundance of insects and other wildlife. May many more hedges be created.'

Vicki Hird, author of *Rebugging the Planet*

'A passionate celebration and exploration of our wonderful hedgerows. You'll be swept along by Hart's intoxicating enthusiasm.'

Dave Goulson, author of *Silent Earth*

'At the Soil Association, we have long advocated for the reinstatement and proper care of that multifaceted boon to nature and farming, the hedgerow. With Christopher Hart's wonderful *Hedgelands* by your side, any farmer, conservationist or interested layperson will understand so much more about this extraordinary resource, and how to appreciate and care for it.'

Helen Browning, chief executive, Soil Association

'Hedgerows are linear forests. Christopher Hart shows us these jewels in the heart of the English countryside and how biodiversity makes them hum with a richness the planet needs. If we pay attention, the nightingales will sing for us, again and again.'

Diana Beresford-Kroeger, author of *To Speak for the Trees*

'*Hedgelands* is a marvellous walk through a traditional European agroforestry system, complete with information on benefits to biodiversity, wild foods, carbon, climate resilience and more. Highly recommended.'

Eric Toensmeier, author of *The Carbon Farming Solution*

HEDGELANDS

HEDGELANDS

A wild wander around
Britain's greatest habitat

Christopher Hart

with Jonathan Thomson

Chelsea Green Publishing
London, UK
White River Junction, Vermont USA

Commissioning Editor: Muna Reyal
Project Manager: Susan Pegg
Copy Editor: Susan Pegg
Proofreader: Caroline Taggart
Indexer: MFE Editorial Services
Page Layout: Laura Jones-Rivera

Printed in the United States of America.
First printing March 2024.
10 9 8 7 6 5 4 3 2 1   24 25 26 27 28

ISBN 978-1-915294-47-0 (US paperback)
ISBN 978-1-915294-58-6 (US ebook)
ISBN 978-1-915294-21-0 (audiobook)

Chelsea Green Publishing
London, UK
White River Junction, Vermont, USA

www.chelseagreen.co.uk

Is there under *Heaven* a more glorious and refreshing object of the kind, than an impregnable *Hedge* of *one hundred* and *sixty foot* in *length*, *seven* foot *high*, and *five* in *diameter*, which I can shew in my poor *Gardens* at any time of the year … It mocks at the rudest assaults of *Weather*, *Beasts*, or *Hedge-breakers*.

John Evelyn, Sylva, Or a Discourse of Forest-Trees

The custom is in this part of Hertfordshire … to leave a *border* round the ploughed part of the fields to bear grass and to make hay from, so that, the grass being now made into hay, every corn [grain] field has a closely mowed grass walk about ten feet wide all round it, between the corn and the hedge. This is most beautiful! The hedges are now full of the shepherd's rose, honeysuckles, and all sorts of wild flowers; so that you are upon a grass walk, with this most beautiful of all flower gardens and shrubberies on your one hand, and with the corn on the other. And thus you go from field to field (on foot or on horseback), the sort of corn, the sort of underwood and timber, the shape and size of the fields, the height of the hedge-rows, all continually varying. Talk of *pleasure-grounds* indeed! … This is a profitable system too; for the ground under hedges bears little corn, and it bears very good grass.

William Cobbett, Rural Rides

Contents

A Hedge in Wiltshire

We are standing looking at a hedgerow in a leafy corner of South West Wiltshire.

There are three of us: Harry James, a bright-eyed young ecologist with a rather hip Viking hairstyle and beard, and a degree in wildlife conservation. His parents run our local village shop.

There's Jonathan Thomson, a rangy New Zealander, born into a family of dairy farmers on the North Island. He subsequently travelled the world, hung out on a beach in Mexico, worked on a fishing boat netting and potting off the west coast of Ireland and then for seven years as a chef in one of Sydney's top restaurants. He's an inexhaustible enthusiast for all things wild.

And there's me: a writer, with less hair than Harry and less experience of being one of those really cool, annoying, globe-trotting, been-everywhere-done-everything-types than Jonathan.

But my wife and I are also lucky enough to own a seven-acre patch of land in a corner of Wiltshire, including about three hundred yards of beautiful, unkempt, pullulating hedgerow.

Along with listening and talking to Jonathan and Harry, and reading countless books and articles, that hedgerow has taught me a great deal over the past few years.

The hedge we are contemplating, however, is Jonathan's, and it's the original inspiration for this book. It stands in a quiet corner of twenty-five acres of rewilded countryside in South West Wiltshire, which Jonathan has created out of a mix of farmland, woodland and bog. Rewilding has taken years of back-breaking but deeply rewarding work, with Harry doing a huge amount as well, along with Jonathan's brother-in-law Patrick and numerous volunteers. I, too, have done my bit, standing around watching them work, making amusing jokes and drinking several cups of tea.

The result is a marvellous mini-Knepp, which Jonathan has christened Underhill Wood Nature Reserve. With its mature trees, open grassland, damp ditches, beautiful lake, heaps of brushwood and countless piles of gently rotting logs – beetle heaven – it bustles and hums with life.

But it is this particular hedge that really caught the attention of Harry the ecologist – because the explosion of wildlife that has taken place here, after it was 'conservation laid', has been truly spectacular.

Our Greatest Achievement

Hedges are one of the happiest accidents in human history. Originally created, several thousand years ago, as livestock barriers or boundary markers for fields freshly cleared from the wild, they rapidly became some of the richest and most favoured habitats of many of our native flora and fauna.

The very first hedges were probably 'dead hedges', built up from the surplus branches and brushwood generated when woodland was cleared. Yet even these would rapidly have been colonised by all the creatures that live in and feed on dead and decaying wood, from fungi upwards. In no time at all, though – as I have witnessed on my own patch – a dead hedge turns into a line of vigorous nettles, thriving on the abundant nutrients released into the soil by the decaying wood.

Meanwhile, the birds are going in to perform one of their numerous ecosystem services, perching awhile on any comfortable-looking branch or twig and absent-mindedly seeding the first generation of living shrubs and trees that will soon grow up from the midst of the dead hedge: the seeds, in other words, are eaten by these random avian horticulturalists, and then pooed out.

Birds like thrushes often swallow haws whole, and blackbirds sloes, digesting the fruit while cunningly avoiding a dose of the hydrogen cyanide from the stone that might poison them by pooing it out untouched. The same goes for crab apples, spindle berries and so on, while jays may bury their acorns and squirrels cache their hazelnuts, all in the shelter of a dead hedge, soon to sprout into a living green line of shrubs.

Our ancestors learned to clip an unruly hedge or one that was becoming gappy; better still, to cut halfway through a rising trunk and then lay it back into the hedge sideways. Thus, the ancient art of hedge-laying was born, eventually creating a barrier so interwoven and strong that it could even hold back a bull wanting to move fields to get to some cows, while simultaneously producing a dense, thorny habitat that stood, and stands to this day, as a perfect proxy for the scrubby thickets that once dotted our primeval wood-pasture landscape.

This is an absolutely crucial aspect of the human-made hedgerow, which I will be doggedly banging on about throughout the book: their uncanny similarity, in ecological function, to the thickets that so much of our wildlife has depended upon for millennia. A thicket may be defined as a densely packed, often near-impenetrable group of low shrubs and trees, which in the wild will be kept in its tight shape by peripheral grazing. If you look at the second picture in the photo section of this book, you will see the landscape now emerging at Knepp, shaped by the grazing and browsing of large herbivores: it's almost *all* thicket – or is it hedge? Impossible to tell the difference, which is just the point.

But as Knepp's presiding genius, Isabella Tree, lamented recently:

> We've forgotten about this precious resource in recent times. In the days before plastics and mass manufacture, thorny scrub provided us with charcoal, medicines, dyes, food, fodder, thatching, furniture, gunpowder and tools. Above all, it was valued as a nursery for trees. 'The thorn is the mother of the oak' is an ancient forestry saying. Thorny scrub was considered so valuable that a 1768 statute in the New Forest imposed three months' forced labour and lashes of the whip on anyone found extracting it. Now, however, we consider scrub 'wasteland' and eradicate it wherever it appears, and even on nature reserves, volunteers spend weekends 'scrub-bashing'.[1]

As well as substituting remarkably well for this precious thorny scrubland, opening onto either arable land or grassy pastures, the hedgerow also replicates the richest bit of any woodland: the woodland edge. No wonder our native hedges are such superb reservoirs of wildlife. From a bird's-eye view – or a weasel's, or a ground beetle's, or a primrose's – they *are* thickets and woodland edges combined, offering perfect habitat, safety from many predators, temperature control on both excessive heat and cold, shelter from the wind, ample food in the form of nectar and pollen, fruit and nuts, and numberless insects and larvae.

If we set out tomorrow to devise some ingenious new form of human-made habitat for our hard-pressed wildlife, we would never equal the wonders of the good old British hedge.

The Hedge Habitat

The traditional British hedge is the greatest edge habitat on earth. It is a green food bank, a windbreak, a stock fence, a flood defence system, an immense storage unit for excess carbon dioxide and an incomparable haven for wildlife. 'Hedges may support up to 80 per cent of our woodland birds, 50 per cent of our mammals and 30 per cent of our butterflies,' says the RSPB.[2]

A hedge provides singing posts for birds, a crucial navigational aid for bats, and a cross-country route for any number of small mammals safely hidden from predators. Hedgerow shrubs and trees, bathed in sunshine, will also produce far more fruit that in a woodland. You won't find many juicy blackberries in the heart of a dark oak forest, no matter how many brambles might be trying to grow there.

The most generous kind of hedge should include shrubs, trees and bushes, coppiced and/or cut and laid, forming a row. Due to their effectiveness in containing livestock, the thorniest of our native shrubs and trees will predominate, such as blackthorn, hawthorn, dog rose and crab apple. And, again, by happy chance, these thorny species are also among the most valuable for wildlife. Hawthorns, for instance, are the equal of beech or sweet chestnut for providing fruit and seeds, while their leaf litter – an often-overlooked micro-environment – is rivalled in importance only by the ash. A unique double.

Above this main scrubby, shrubby line that is the classic hedgerow will, ideally, tower some fine mature trees, with some showing signs of decay: not too many, though – one every few dozen yards is ideal. And then cladding the sides of the hedgerow all the way along, spilling down to ground level, will be ivy, bindweed and brambles, and then smaller flowers like celandines and dandelions, bluebells and perhaps even orchids, as with the Underhill hedge (see the central picture on page five of the photo section). From end on, then, the shape of an ideal hedgerow will be a sprawling, spiky and messy pyramid, with a wide and generous bushy base (see the bottom photo on the aforementioned page). If of sufficient age, the entire hedge may sit on top of a venerable earthen bank, offering a whole new range of home-making possibilities to burrowing rabbits, badgers, stoats and so on, while the very best hedge will also feature a good damp ditch.

Taken together, such a hedge forms a narrow but incredibly complex ecosystem in a small area, comprising elements of mature woodland, woodland edge, thicket, flower-rich grassland and even wetland, which alone explains why it is so incredibly valuable a feature of our countryside.

Hedge Cuts

We will go into more detail about hedge-laying in chapter seven, but for now, the description given by the National Environment Research Council on how they cut a conservation hedge is very clear:

• Stems were cut at the base as for traditional Midland-style hedge-laying. Less woody volume was removed. Remaining stems and branches were laid along both sides of the hedge. Stakes were used sparingly and no top binding was used.

- Three years after rejuvenation, hedge structure and rates of woody re-growth were as good as traditional hedge-laying in improving hedge condition.

- Berry provision for overwintering wildlife was slightly better than for traditional hedge-laying immediately following rejuvenation.[3]

In simpler terms still, conservation hedge-laying is faster and less precise than the traditional laid hedge: the spikier, scruffier and wider a hedge is, the more wildlife loves it. While the stem is cut just the same as in any other hedge-laying technique, no side shoots are removed and the entire stem is then simply shoved over – it can feel quite brutal – and laid down on top of or even alongside the previous one.

Such a hedge will be a nature-lover's delight: humming with bees and hoverflies, bright with ladybirds and delicate lacewings, filled with springtime birdsong, and with buds and blossoms; wreathed with climbers like honeysuckle and bryony in summer; or spilling out in abundant falls of red and golden leaves in autumn, with browning hazelnuts, scarlet berries of the guelder rose and hawthorn, and the blue-black clouded sloes. This practical and time-honoured stock barrier is also one of humanity's most beautiful creations.

Along a stretch of just thirty yards or so of flourishing conservation-laid hedgerow, especially if it includes wide grassy margins (as it should), you will find dozens of species of flowers, woody shrubs and very likely a mature tree some 200 to 300 years old, perhaps pollarded some time in the past. Unseen within the hedge, there will be a busy corridor of mammals like shrews, wood mice and bank voles, as well as reptiles and amphibians, particularly in hedges that are accompanied by a damp ditch.

Coming back to our specific prize hedge at Underhill, it really wasn't that difficult to restore what had been a straggly, grown-out hedge to its former glory: a mighty burgeoning barricade, twice the height of a man and almost as deep from front to back; festooned with buds and flowers in springtime; in summer a tumultuous carnival of insect life; in autumn a dazzling spread of numberless fruits and nuts, first come first served; and in winter a precious haven for hibernating animals or fluff-feathered birds, sheltering from the silent frost and the long, iron-dark night.

Certainly, in terms of labour, creating or recreating such a hedge was nothing like as demanding as digging out a lake or pursuing an extensive new tree-planting programme. It simply involved Jonathan and Harry donning their thick leather gauntlets, taking up their saws, axes and billhooks, and setting to work on this relict and wind-blown structure of rapidly diminishing ecological value.

By relaying it over the course of a couple of days – nicking, sawing and bending over existing trees and shrubs with deep and powerful roots and therefore a great capacity for renewed growth – they achieved remarkable results. Crucially, they realised, this endeavour could easily be replicated across vast swathes of Britain, to the almost immeasurable benefit of wildlife and the health of our ecosystem.

By just the following winter, Jonathan was already beginning to sense that the effect of relaying this particular hedge was immense. He describes how he stood beside the dense structure one winter's day and heard a 'mad cacophony' of birds within. The din was caused by those two noisy, much-loved species of the thrush family, the redwing and the fieldfare, winter visitors to our islands that often flock together.

The redwing, the smallest of our native thrushes, can be identified by – you guessed it – the red patch on its wings, while the fieldfare is a larger bird with a distinctive grey head and tail. Arriving from Scandinavia in winter to escape the northern cold, large numbers of them can be seen in dense, traditional-style hedgerows and fields, often looking for their favourite high-energy food: wild berries. They make a fantastic racket, too, as they gather in their 'straggling, chuckling flocks', as the RSPB pleasingly puts it. But, like so many of our native birds, redwings are on the Amber List, while fieldfares are on the Red List, denoting that they are 'in critical decline'.

Yet the experiment at Underhill suggests that this could be reversed with one simple measure: conservation hedge-laying.

Previously, says Jonathan, species like fieldfares and red-wings had been passing visitors to Underhill, but this winter they had actually moved into the hedge. It was already close-packed, impenetrable to most predators, especially corvids; one study found that a conservation hedge could reduce corvid predation by an astonishing 84 per cent. It was also a wonderful shelter from the wind and the cold, and absolutely festooned with red and purple fruit: hips and haws, sloes and blackberries, and the rest.

Harry, for his wildlife conservation degree dissertation, 'Hedging Your Bets: The Effects of Conservation Hedgelaying on Invertebrate Assemblages in Hedgerows', then carried out a much more detailed and technical survey over the next four years, measuring the change in invertebrates in the hedge, and discovered a relentless increase in biodiversity that was deeply impressive: 'Species richness was seen to rise by an average of 2.5 species 1–2 years after initial management and 6.7 species after 3–4 years.'

He also found individual numbers, or 'abundance', of invertebrates rose by some 20 per cent over one to two years, and 40 per cent over three to four years. Some of these species included such insects as click beetles, Darwin wasps, lacewings and sawflies as well as orb spiders, wolf spiders and comb-footed spiders.

This is an incredibly quick result in ecological terms. But Harry and Jonathan were also fascinated to think that if this rate of increase were to continue, as they suspected it would, a conservation-laid hedge might see these numbers increase by 100 per cent, or double, in just ten years.

Harry also noted that the 'foliage density' increased from 20 per cent to 100 per cent in just four years. It quintupled, in other words.

Although it was beyond the remit of his dissertation, it is obvious that such a huge increase in foliage, and of a 'building block' group of animals like the invertebrates, would also have hugely beneficial effects on everything from carbon capture to the creation of rich habitat for birds, reptiles and mammals – not to mention on the pollination of trees, flowers and crops.

It's particularly important to emphasise that these impressive results were achieved over a period of just four years. As we enter a time of real ecological urgency, we are going to need fast solutions to many of our problems – faster than nature, in many cases, can accommodate. Certainly, planting vast new stretches of young woodland, which seems to be the favourite green strategy at the moment, achieves very little indeed within such a timeframe. The great Northern Forest is one example – an ambitious plan to plant some fifty million trees across ten thousand miles of the UK, from Liverpool eastwards to Hull. Thirteen million trees have already been planted, and serious questions have been raised about the point of this 120-mile-long plantation, since very young, dense-canopy woodland is of low ecological value. It will, admittedly, help to produce cleaner air and mitigate flooding, but is it really the best approach, and the most useful way to spend £500 million?

We'd be much better off looking after our existing ancient woodlands – and our hedges. As the great Dr Oliver Rackham, professor of historical ecology at Cambridge and probably the foremost expert on Britain's woodlands, said years ago, planting trees is 'an admission that conservation has failed'. The aim should be to preserve the superb ancient trees that we have – and the hedgerows in which many of them still grow.

The Modern Alternative

But the typical twenty-first-century farmland hedge very rarely looks like the relaid, fruit-covered, bird-crowded Underhill hedge. We are losing an incredibly precious and self-sustaining feature of our countryside. According to the UK Centre for Ecology and Hydrology's 2024 'Land Cover Plus: Hedgerows' report England alone has 390,000 kilometres of hedgerows between 1 and 6 metres high, plus around 250,000 kilometres of lines of vegetation that include degraded and overgrown hedgerows, which would barely be recognisable as hedgerows at all. The proportion of neglected, decayed and relict hedges, alas, is growing all the time. Other estimates put the total figure at more like 400,000 kilometres – it's a famously difficult thing to assess – and we will go with this figure in this book. On a happier note, there are of course many more splendid hedgerows in Wales and parts of Scotland, as well as a hugely impressive 400,000 kilometres in the Republic of Ireland.

But what does this really mean? You don't need to travel far in the English countryside to see the sadly typical modern hedgerow of agro-industrial farmland: a miserable, intermittently gappy line of stunted blackthorn and hawthorn, with little or no grassy, flower-rich margins whatever. If such a deteriorated hedge is on arable land, it may then be regarded by the farmer as nothing but an obstacle to his tractor, and removed entirely, something which has happened on a devastatingly large scale across the grainfields of East Anglia. Farmers there are only now beginning to learn – or relearn, rather, as their farming forebears understood it pretty well – how the elimination of their hedgerows has meant ever-increasing soil erosion and falling crop yields.

If the neglected or harshly flailed hedge is on a sheep, dairy or beef farm, it will seem much cheaper for the farmer to simply install a new barbed-wire fence just in front of the hedge. You can see this in every corner of Britain today.

A barbed-wire fence may be cheaper in the short term – but it may be much more expensive in the long term, as we shall be exploring later in this book. We will also be looking at how farmers – far from the cartoonish villains of the countryside, but, rather, hard-pressed food producers just trying to stay in business – might be incentivised to forego the barbed wire and return to protecting their hedges instead.

A conservation-laid hedge will pay us back a hundredfold, and from numerous different perspectives, as subsequent chapters will demonstrate. From plant life, insect abundance, homes and sustenance for birds and mammals, to providing windbreaks, flood defences, barriers to soil erosion and climate change, to (my personal favourite) offering us year-round and almost limitless Food for Free (as the title of Richard Mabey's famous book has it), the benefits of the ecologically ideal British hedgerow are almost countless.

But the blunt truth is that, as things stand in the British countryside, the typical hedgerow looks simply… trashed. Or dying.

Hedges are of course artificial, human-made creations, and once established they need continuing human management to flourish.

If nothing is done to them at all, the individual trees and shrubs that compose them simply keep on growing taller and taller, until eventually you have a line of gappy, slightly straggly trees – as was happening to the Underhill hedge before relaying. Although not worthless in ecological terms, this is no longer any good as a stock barrier or windbreak, and nothing like as valuable a wildlife habitat as a dense, tangled, thickety centuries-old hedgerow.

But if today's farmland hedgerow *is* managed, the chances are that this simply means an annual flailing in autumn with a powerful machine drawn by a tractor, chopping and slicing the structure down to a neat oblong box shape, shorn of all fruit and leaving a hedge that is a 'degraded habitat', as Harry puts it. It will be composed mostly, in the ghastly phrase of the conservation charity Buglife, of 'a mass of scar tissue'.

'Repeated flail cutting at the same height will eventually produce a mass of scar tissue and dead branch ends which support few healthy shoots. If the flail cut is taken back to the main upright trunk of hedges, the bark can be torn off.'[4]

Anyone who has passed by a recently flailed hedge will know well the smashed wood and heaps of splinters it produces – not to mention the feeling of wintry lifelessness. Farmers might say that it is 'cheaper' this way than traditional hedge-laying by hand, at a rate of some thirty yards per day per person. But have the wider costs and benefits really been fully calculated?

Cutting hedgerows by heavy modern tractor and flailing machine brings deafening noise, soil compaction and burns diesel fuel at approximately a gallon per hour, while cutting the autumn hedgerows and destroying their fruit harvest seems frankly bonkers. Yet farmers under current rules are allowed to start flailing from 1 September, and many do so.

The heady abundance of red and purple fruits thus destroyed aren't lost only to wildlife but to people as well. Long before terms like 'forest gardening' and 'permaculture' came into use, the old British hedgerow was already offering such things. But we continue to destroy a vast crop of hedgerow fruits, nuts and greens annually, at considerable expense, and then airfreight in fruit from around the globe. Aldi currently airfreights in strawberries from Egypt and blackberries from Peru.

And some people still say it's environmentalists who are delusional!

The Future

Hedges are far more than quaint relics of the old-fashioned countryside that should either be left to quietly die out to make things more 'efficient' or else be flailed to within an inch of their lives. They form an unbroken line back through history, quite literally in some cases, to the native scrub of our ancient landscape. The celebrated 'Judith's Hedge' in Cambridgeshire, for instance, is at least 900 years old – older than either Windsor Castle or Westminster Abbey, first sprouting at about the time the Crusaders were capturing Tyre.

At the same time, carefully nurtured hedgerows hold out a promise for the future as few other features of our landscape do. In a time of increasing food scarcity and supply-chain disruptions, a richly flourishing hedgerow can offer an abundance of free wild food, from greens in the spring to fruit and nuts in the autumn, as well as measurably increasing yields in the field it protects.

In a time of increasing rainfall, flood risk and soil erosion, hedges form a resilient living barrier to such challenges. And in a time of deeply destabilising and unpredictable climate change, they offer not just a vague, romantic sense of stability and continuity with the rose-tinted past, but a literal stability, locking up megatons of carbon dioxide in their complex and artful structures.

The relaid hedge at Underhill that first fascinated us, as it roared back to abundant life, seems a template for how many thousands of miles of hedgerow across England might look. With a small initial investment of time, energy and craftsman-like skill, they will repay huge dividends. The campaign group Rewilding Britain has stated their ambitious aim of seeing fully a third of our native hedgerows renewed as conservation hedgerows. This would create a stunning living latticework of some 150,000 kilometres of conservation-laid hedgerow across our countryside. And, unlike some rewilding projects or tree-planting programmes, which can cause controversy by taking food-producing farmland out of production, the restoration of our existing hedgerows merely means better management of these fabulous linear ecosystems already in place, and uses up no extra land whatsoever.

It is hard to imagine a greater monument to our faith in the future.

But hedges need maintaining, and a truly wildlife-friendly hedge needs to be laid, by hand, so that it can continue to grow and thicken in perpetuity. Where this has been done, we now still have flourishing hedgerows that go back certainly to Anglo-Saxon times, and arguably to the Bronze Age.

At first glance, the cost of hand-laying all the hedges on our country's farmland seems so prohibitively expensive as to be a mere pipe dream. But costing such things is more complex than it looks. Farmers currently flail their hedgerows every year, whereas a well-laid hedge only needs laying once every ten years – and was sometimes even done on a *forty-year* cycle – resulting in colossal cathedral-like hedgerows in the meantime.

There are also the vast environmental benefits of flourishing ancient hedgerows, and their ecological services from flood control, soil protection and climate moderation, to the conservation of innumerable species of hedgerow-dwelling insects – especially at a time of devastating invertebrate decline.

With better management of at least a good proportion, if not all, of our incredible half a million kilometres of hedgerow, we might not need to worry so much about insects disappearing, bird numbers falling or our targets for carbon capture. Our lovely native hedgerows would do much of the work for us, if we only looked after them.

These accidental nature reserves stitch our countryside together into that famous 'patchwork quilt' that can be seen from virtually any hillside, moorland or downland in England or Wales. But when the stitching starts to fray and fail, and the patchwork quilt starts to come apart… What will we have to keep us warm when winter comes?

It would be wise to remember the old saw: a stitch in time saves nine. We should be saving and renewing our ancient hedgerows with the same zeal as our ancient woodlands. They are just as valuable.

It isn't just farmland that has hedgerows, of course. There are some twenty-three million gardens in the UK, the majority of which will have at least a few yards of hedge. There are more hedgerows around allotments, public parks, school playing fields, churchyards, hospital grounds, university campuses…

This is an incredible potential resource for wildlife, yet many are machine-clipped, monocultural privet or beech hedges. It would be far more imaginative and fun, and produce much greater sensory pleasure, to 'rewild' a boring suburban hedge by introducing delightful flowering climbers like honeysuckle and bindweed, or even, if you really wanted to annoy the tidy-minded, a few flower- and fruit-rich brambles.

And if all such non-farmland hedges were supplanted wherever possible by the richly diverse and conservation-laid kind, and allowed to grow higher, thicker, thornier, with wide bases and wide grassy verges mown only once a year, late in summer, like mini hay meadows, the effects might be immense.

Given recent estimates that invertebrate numbers worldwide have declined by as much as 75 per cent in the past fifty years, the ecological benefits of simply conserving and expanding our existing network of hedges, on both agricultural and amenity land, could be tremendous.[5] Invertebrates are one of the foundation stones of a healthy ecosystem, insect declines being a major cause of similar declines in bird and mammal populations.

This is confirmed at Knepp, where allowing hedges to bush out to as much as seven metres wide, and the subsequent leap in insect numbers, has been central to the return of the nightingale and cuckoo there, and to the rise in recorded bat species from five to thirteen in just a few years.

Hedges have always been greatly valued for their usefulness in demarcating land as well as proving an endless bounty of free food and healing herbs, but they can also denote something a bit rough and ready. Happily, that's exactly how we think a hedge *ought* to be: not neat and clipped, polite and suburban; but as rough and tangled, thorny and thicket, outrageously sprawling and overgrown as possible. Well-mannered hedges are boring – and wildlife doesn't like them either. Wildlife, it comes as no surprise to learn, likes its hedges *wild*.

As we progress though this book, we will examine the wonder of the hedgerow from many different angles: the history of the hedge, hedgerows and insects, hedgerows and foraging, and so on. But in every instance, it will be clear that if what we have learned at Underhill, from a single closely studied example, could be applied on a much wider scale across both our urban and rural areas, it would vastly improve our environment and the richness of all our lives – not to mention help stabilise our climate, preserve species diversity, provide us with the finest free food at virtually zero effort or expenditure, and help to ensure a common future for us all.

And, ironically, as we shall see next, many of the natural benefits of the hedge were known about in the past, through tradition and custom rather than cutting-edge environmental science.

Hedge-Priests, Hedge-Taverns and Hedge-Mumpers

A Brief History of the English Hedge

*MUMPER.— English synonyms. Abram-man
(or -cove); bawdy-basket; milestone-monger;
mugger; ruffler; scoldrum.*

John S. Farmer and William E. Henley,
Slang and its Analogues, Past and Present

Before going into more depth with the natural history and wondrous ecology of the British hedgerow, we will begin with its place in our history. And one of the best ways to understand how people and hedges have interacted over time is by looking more closely at the English language itself.

The Word 'Hedge'

The extraordinary richness of terms relating to hedges, many now obsolete but highly entertaining, and still to be found in old books and dictionaries, remind us of the centrality of these living boundaries in days gone by.

For instance, the term 'hedge' was often affixed to another noun to denote something slightly wild, backcountry and uncivilised. The indispensable *Brewer's Dictionary of Phrase and Fable* explains: 'The use of hedge for vagabond, or very inferior, is common; as hedge-mustard, hedge-writer (a Grub Street author), hedge-marriage (a clandestine one), etc. Shakespeare uses the phrase, "hedge-born swain" as the very opposite of "gentle blood." (1 *Henry VI.*, iv. 1.)'

Hence the stream of lively insults in Ben Jonson's *Bartholomew Fair* (1614): 'Out, you rogue, you hedge-bird, you pimp, you pannier-man's bastard, you!'

It was urban snobbery and citified condescension, in other words, of the same variety that labelled religious outliers and nonconformists 'pagans' and 'heathens', which, strictly speaking, denote simply 'rustics' and 'heath dwellers' respectively. Today's equivalent might be calling country people local yokels, hicks or rednecks.

'Hedge-priests' were wandering and (allegedly) ill-educated priests, possibly ones that had been defrocked for unclerical behaviour. Alternatively, they may have failed to switch allegiances with sufficient smoothness when the powers that be suddenly decided we were Protestant, or Catholic again, or Puritan, or Episcopalian – which is why the word suddenly comes into wide circulation in 1660, the year of Charles II's restoration, when so many dogged Puritan clergymen were rudely ejected from their parishes.

The poor but principled parson might continue to scrape a living by offering 'hedge-marriages', as John S. Farmer and William E. Henley's 1893 *Slang and Its Analogues Past and Present, Volume 3* explains, solemnised by the ancient if somewhat unorthodox method of the happy couple holding hands and jumping over a broomstick. This ceremony might be particularly urgent in the case of any young country lass who had been 'hedge-docked' – or, as the editors delicately explain: '(venery).— Deflowered in the open.' – and was now expecting a 'hedge-brat' that had been 'hedge-begot'.

The hedge-priest par excellence must surely be John Ball, the turbulent cleric at the heart of the Peasants' Revolt in 1381. He wandered abroad without a parish, preached fiery sermons (mostly in Essex) attacking the gluttonous ruling class and advocating social equality, frequently clashed with the Archbishop of Canterbury and was eventually hanged, drawn and quartered for his troubles. He is memorably characterised in *A Dream of John Ball* by the great Pre-Raphaelite artist and writer William Morris, at one point half-wishing that he had learned to be 'a trembling cur to creep and crawl before abbot and bishop and baron and bailiff', rather than noticing 'the evil of the world wherewith I, John Ball, the rascal hedge-priest, had fought and striven in the Fellowship of the saints in heaven, and poor men upon earth.'

Non-clerical gentlemen of the road – i.e. tramps, also known as hedge-mumpers – might also take up overnight residence in the shelter of a hedge. Farmer and Henley again: 'The vagrant brotherhood have several terms for sleeping out in a field or meadow. It is called "snoozing in Hedge Square," etc.' Homeless vagabonds clearly loved to make light of their situation with comically elaborate euphemisms; French tramps similarly talked about *coucher à l'hôtel de la belle étoile* and *coucher dans le lit aux pois verts*, which mean 'sleeping at the hotel of the lovely stars' and 'sleeping in a bed of green peas', respectively.

Perhaps the itinerant 'hedge-mumper' would have spent his evening in a 'hedge-tavern', a dismissive phrase dating from at least the seventeenth century. In the mysterious gentleman B.E.'s

A New Dictionary of the Terms Ancient and Modern of the Canting Crew of 1690, a hedge-tavern is defined as 'a Jilting, Sharping Tavern, or Blind *Ale house*', which doesn't exactly help; but, again, the suffix 'hedge' here denotes something dodgy, dirty and not the sort of place that well-born city folk would go for a pint.

The sense of hedges as places of resistance and unruliness is still more evident in the mainly Irish term of 'hedge-school'. In the long tragicomedy of Anglo-Irish relations, the nineteenth century saw Penal Laws imposed by Westminster that forbade any but Anglican children from attending school – thereby depriving the vast numbers of Irish Catholic children of an education. In response, Irish teachers and priests set up secret hedge-schools, also known in Gaelic as wayside schools, to teach history, Latin and Greek, and the Catholic faith. Though they wouldn't always have been located in actual hedges, any more than were hedge-taverns – the schools would more likely be found in barns – the word again suggested their illicit and clandestine nature.

Like the countryside generally, hedges were often held in suspicion as places of unseemly liberty and resistance, just as much as they were valued as boundary markers and stock barriers for wealthy landowners. They were like miniature 'greenwoods' of the kind where Robin Hood and his men hid out and fought back against Norman tyranny.

It's all rather contradictory and revealing. A parallel usage would be the word 'dog': man's best friend, protector of flocks, a loyal and faithful companion – yet a word also applied to anything deemed inferior, such as the poisonous dog's mercury, the unscented dog violet or the lethargic, bad-tempered dog days of late summer.

The language of place names – the lovely science of toponymy – leads us even further back into our rural past.

Glimpses of our hedge-rich history can be found in names such as Hagley in Worcestershire, meaning hawthorn meadow or woodland clearing where the hawthorns (hags) grow; and Heigham Potter in Norfolk, the 'potter' element denoting simply that a potter lived there, and Heigham, or 'Echam' in the Domesday Book of 1086, meaning 'homestead or enclosure with a hedge'. Haworth means just the same, while Haywards Heath means 'heath by the enclosure with a hedge'.

And one can't help feel somewhat haunted by the ghosts of our islands' wilder and more beautiful past, concealed beneath the modern names of the most urbanised and unpicturesque of places: Croydon, meaning 'the valley where the wild saffron grows', or Battersea, 'the island in the marsh belonging to a man called Beaduric'.

The Earliest Field Boundaries

Our ancestors knew all about the value of field boundaries and hedges thousands of years ago and did a great deal of work to create and maintain them.

In County Mayo, at the Céide Fields, there are dry stone walls covering some one thousand hectares, around what would once have been wheat and barley fields that date from 3000–3500 BCE – the Neolithic – created about the same time as the astonishing Newgrange was built in County Meath, and far older than the Pyramids. Yet these Stone Age field patterns 'do not strike the eye as being exceptionally archaic or unusual', says hedgerow expert Richard Muir, showing how little the essential idea of a field and boundary has changed in five thousand years.[1] Such amazing continuity also explains why they are so loved by our wildlife: because they've had so long to get used to them.

The field systems of Dartmoor are ancient too, the famous Grimspound being Bronze Age, from around 1300 BCE – and amazingly atmospheric a place it is too, especially if you visit it on a day when the fog is rolling in across the moor from the west… It has been variously interpreted as a Druid temple and even as a defensive group of huts for young bachelor warriors. But being at the centre of a vast 'reave' or field system, that is to say, a pattern of clearly established and demarcated fields, it looks much like a well-defended farming settlement with cattle corral to me. And such human-made stock barriers were already becoming havens for wildlife, just as planted and living hedgerows would do later.

There are more Bronze Age field systems in the far west of Cornwall and on the Isle of Purbeck in Dorset, where pollen remains show they were growing emmer wheat, barley and beans by 1400 BCE. Oliver Rackham has pointed out that the stone-clad, often gorse-topped hedge banks of Cornwall may be among the oldest human artefacts on the planet still in use for their original purpose. More recent Celtic and Iron Age field patterns, from around 500 BCE, are yet more obvious: they tend to be long and narrow, angled against the prevailing south-west winds.

It isn't too difficult to see how and why these boundaries arose. For ploughing good but stony soil with a wooden ard or ploughshare, it would be much better to clear out the stones first – a job no doubt for children. You can still find fields in the English countryside today with piles of stones at the side as a result. But if you had enough stones, you could use them to build walls around your field, thus reducing wind speed and soil erosion, and sheltering your crops.

Knowing all about rotation long before the Agricultural Revolution of the eighteenth century, as well as windbreaks and soil erosion, our sensible Bronze Age forebears regularly planted their beans to fix fresh nitrogen and 'folded' their sheep on the grain fields; that is, they brought them down from the high downlands and grasslands at night to eat any weeds and dung the soil. Then the stone walls served as useful livestock barriers.

All of these primarily stone structures were also fabulous havens of wildlife, especially for lizards and snakes. It's just as useful to leave a pile of stones as a pile of logs in your garden. But are they really hedges? Not yet. We take a hedge to be a living human-made barrier, the etymology of the word itself showing a connection to the Old English word for the hawthorn: the hedging tree par excellence.

But first would have come dead hedges, where fields and clearings were created from woodland by 'assarting' (converting forest to arable land), resulting in vast piles of cut timber and brushwood or brash. This would have been immediately handy to create dead hedges. Excavations at the amazingly well-preserved Must Farm site at Whittlesey, Cambridgeshire, revealed a seventy-metre dead hedge with a row of coppiced stakes, dating from 2200–1950 BCE.[2] So, the art of dead-hedging in this country is *at least* four thousand years old. And, after a hard day's work dead-hedging, we also know that the residents of Must Farm liked a good dinner of wild boar, venison, pike and wheat porridge.

Yet if we look worldwide, then dead hedges are far, far older. As a barrier to keep animals *out*, rather than in – such as hyenas, lions and so on – they may be as ancient as we are as a species. Hunting parties out on the plains would have thrown up overnight thorn barriers to keep wild animals away from their campfire. A 2013 study by Georg Müller in *Europe's Field Boundaries* found evidence of dead hedges being used far back in the Stone Age and live, shrubby hedges by 11,000 BCE.

But there is much speculation among landscape archaeologists about if and when early farmers in Britain deliberately planted and then managed and laid the first live hedges – although, as mentioned in the introduction, you really don't need to because, thanks to birds eating and defecating some seeds whole, a dead hedge quickly and rather miraculously becomes a live hedge anyway. I estimate that within just ten years you'd have the beginnings of a live hedge in its place with no extra human effort at all. And this, I suggest, is often how our first hedges came about.

Today, we're so used to planting new hedgerows with endless lines of bare-root hedge plants bought from commercial suppliers, and then protecting each one with those beastly plastic tree guards – but perhaps our ancestors, as in so many things, took it slower and got better results for less effort, letting nature do most of the work for them.

In time, they no doubt learned that if you cut and lay a hedgerow you get still better results, and instead of a dead hedge, which might last a year or so before starting to fall apart, a live hedge will effectively last indefinitely – and for much the same reasons that a coppiced hazel will live not just eighty years or so, like standard hazel, but many centuries. Hedgerow trees and shrubs respond vigorously to being cut, and regrow with renewed determination.

The First Laid Hedges

When is the first definite evidence of deliberate hedge-laying in Britain?

Incredibly, the farmer and archaeologist Francis Pryor, an expert on the fen country, has found evidence of hedge-laying dating back 4,500 years – actually older than the first known dead hedge from Must Farm. This is a bit of blackthorn, cut at a deliberate right-angle, found among bundles of twigs in a field boundary ditch in Fengate and preserved in the anoxic soil ever since.

The natives of Britain and Ireland have clearly been hedge-laying for some time. And they were certainly doing so by the Iron Age. An archaeological dig at the Bar Hill Fort, on the Antonine Wall in Scotland, dating from soon after 140 CE, offered evidence of this. The fort was built on a site with pre-existing ditches that had to be filled in and, again thanks to the acid soil, this brushwood infill has been preserved at the site, along with charming items like a child's shoe.

The brushwood consisted largely of hawthorn branches, 'two to six years old … scarred and bent', in precisely the way you would expect if they were cut from a laid hedge and not in any way suggestive of browsing by animals.[3]

And if the archaeology doesn't seem conclusive enough, how about the word of Julius Caesar, no less? For no book about hedge-rows is complete without mention of his famous description, in *Gallic War*, of his campaign against the Nervii, a Celtic/Belgic tribe who occupied modern-day Belgium, and their defences.

'They had succeeded in making hedges that were almost like walls, by cutting into saplings, bending them over, and intertwining thorns and brambles among the dense shoots and branches. These hedges provide such protection that it was impossible to see through them let alone penetrate them.'[4]

Caesar claimed that the Nervii had put these hedges in place to obstruct Roman cavalry, which seems highly unlikely: they were probably already established farmland boundaries, which proved just as troublesome to Caesar's legions as did the mighty hedges of Normandy to the Allies after D-Day in the bocage country.

These ancient and towering hedgerows, up to five metres tall and as deep as dense thorn thickets, even proved effective barriers against tanks, with the US army having to fit cutting equipment to the front of their thirty-tonne Shermans just to get through.

Two thousand years earlier, quite possibly Caesar was only trying to make his own, eventually successful campaign against the Gallic tribes appear more impressive by exaggerating their defences, which were really only field boundaries. You can never quite trust a man who refers to himself throughout his book in the third person.

But certainly the old-established and well-laid hedgerow was known by his day across much of Britain and France, and in original Roman lands too. Farmer writers like Columella and Palladius Rutilius wrote about their usefulness, and Cato the Elder (234–149 BCE) recommended hedges of elm and poplar, stock barriers that would also produce both browsing materials for animals and timber.

Like Cato, I've watched the cows in our field supplementing their standard grass diet with browsing from the hedgerows, without damaging them: in fact, scientists now think that compounds in both cow and deer saliva actually stimulate a browsed tree or bush to grow back faster.

Our cows have eaten oak, elm, willow, hazel, hawthorn, blackthorn, crab apple, wild garlic, hogweed, nettles... The list is endless. They've never come to any harm and are now as fat as barrels and as glossy as butter. As an added benefit, good thick hedges make it a third less likely that your cattle will get TB. As Keggie Carew explains in her brilliant book *Beastly*, it's because wild animals like badgers, who may transmit the disease via their droppings, are much more likely to poo in the tranquil privacy of a dense hedge, at a safe remove from the grass grazed by the cows.

How many farmers today include in their hedge-maintenance plans the calculation that it will also provide fodder and variety for their cattle and sheep, as did old Cato? Not many, I fear. Yet several items on that list have enormous health benefits (though not my trousers), from the antiviral wild garlic to the antipyretic (fever-reducing) willow leaves to the anthelmintic (anti-worms) birdsfoot trefoil. Are cows free to browse on wildflowers and hedgerows healthier than cows confined by barbed-wire fences to boring old rye grass? Almost certainly, hence the growing preference for grazing animals on 'herbal leys', meaning pasture made up of grasses combined with herbs like yarrow and chicory, and legumes like clover and lucerne. But the dietary contribution of the surrounding hedgerows should also be admitted.

Hedges in the Middle Ages

Coming up to more recent times, the Campaign to Protect Rural England (CPRE) report *England's Hedgerows: Don't Cut Them Out!* (2010), gives a clear overview of what we know now of hedgerow history.[5] There's a common idea that our hedges date from the Enclosure Acts, especially from between 1750 and 1850. In other words, as someone has in fact said to me, we'll just be going back to how things were before the Georgian era. What's wrong with that?

The implication is that the disappearance of our precious hedgerows is simply another kind of change in our ever-changing countryside and nothing to worry about. The same sort of argument is heard in relation to climate change: we'll just be going back to the Medieval Warm Period – or perhaps the Eocene. Nothing to worry about.

Let's not get into climate change, but the point about hedgerows is that: a) they have been here far longer than that; and

b) they have become crucial substitutes for the abundant thorn scrub and thicket of our lost wood pastures.

Many of our hedges are far older than the Georgian era, often medieval or Anglo-Saxon, and some two-thirds of England has been a hedge-enriched landscape for a thousand or more years – in places right back to prehistoric times. 'It is only in the Midlands and part of the North East,' explains the CPRE, 'that the majority of these early hedgerows were removed in the Middle Ages to create open field systems, and new hedgerows subsequently planted under the Enclosure Acts.'

In counties that have seen far less change, like rolling, rainy, rich green grassy Devon – sheep and cattle country, so inimical to being ploughed – at least a quarter of hedges date back to the Normans, as old as Devon's oldest parish churches, 'and some are underlaid by banks built in Bronze Age times 4,000 years ago.' (CPRE.)

From the heart of Anglo-Saxon England, we have some remarkably vivid pen portraits of the countryside of the time, thanks to some otherwise pretty dry legal documents known as charters, which gave legal force to certain boundaries.

For instance, a grant of land was made by King Æthelred II – the famously 'Unready' one – of one and a half hides of land (very roughly one hundred acres) at Wetmore in Staffordshire, to Abbot Wulfgeat of Burton Abbey in 1012. The parcel of land was defined thus:

> First, from the Trent where the thieves hang in the middle of the barley meadow, straight on to a spot five field strips before the stockade of Burton, from the strip to the stream, up the ridge to the boundary thorn, then to a ploughland then a long a hedge to a brook, along the brook to a dyke.[6]

Oliver Rackham counted a total of 372 hedges mentioned in surviving charters from Anglo-Saxon England, often referred to by the word *cwichege*, '*cwic*' as in quickthorn, or hawthorn – still the favoured tree of hedges today; and no fewer than forty-eight place names related to hedges. And as Richard Muir also points out, 'the meadow where the thieves hung near Burton in Saxon times is still known as Gallows Flat', again illustrating the extraordinarily long continuity of our landscape history.[7]

An unmistakably luxuriant and very English-sounding hedge makes an appearance in the jolly Middle English poem *The Owl and the Nightingale*, dating from about the reign of Henry II, and probably written by a clergyman called Nicholas of Guildford. The two birds squabble endlessly, being temperamental opposites, the owl being sober and solemn, the nightingale blithe and gay and music-loving and, according to the poem at least, terminally lecherous. It's poignant to realise how nightingales must have been so common then that everyone hearing the poem would have understood the characterisation. Today, these amazing birds are down to about five thousand breeding pairs in the entire country, and few people have ever heard their beautiful song.

But most important of all for our purposes is the poem's description of the nightingale's preferred habitat: a 'vaste thicke hegge'. Exactly the sort of hedge nightingales favour to this day, although all too rare in our machine-trimmed and over-managed countryside. In fact, the medieval poet's 'vaste thicke hegge' sounds like an exact forerunner of the famous 'nightingale hedge' at the Knepp Estate in Sussex.

The nightingales of Knepp are now breeding there in ever-rising numbers, their favoured habitat being – surprise, surprise – 'overgrown hedges 8–14 metres deep, made up of mixed shrub species with a preponderance of blackthorn. The cathedral-like interior of these thorny fastnesses provides the adults and their fledgling chicks with a safe place to forage for insects in the leaf-litter.'[8]

As with so many other of our vanishing flora and fauna, if we want nightingales to recover their previously healthy and viable numbers, all we have to do is get medieval with our hedges, let them grow 'vaste' and 'thicke' once more, and nature will no doubt do the rest.

The medievals were proud and protective of their mighty hedgerows, too, and you could be punished for cutting or damaging another man's. In 1443, an Essex man named James Mede complained that:

> John Palmer senior in the month of March cut down to the ground, took and carried away divers Trees … viz. oak, ash, maples, white thorn and black, lately growing in a certain hedge … and had been repeating this trespass from time to time for 7 years … by which the said James … has suffered damage to the value of 20s.[9]

A skilled craftsman or builder then earned 6 pence a day, or say £100 in today's money, so 20 shillings is equal to about £4,000 – quite a claim!

Hedges were also deliberately created and maintained around young woodland coppice (valuable in so many ways to the medieval economy), which would otherwise have been browsed by deer. We even have a costing from the fifteenth century, for hedge-laying around Bradley Wood near Huddersfield: in the winter of 1457–58, for a length of 1,650 yards, the cost was 41 shillings 8 pence. Again, based on a labourer's day wage of sixpence, you could estimate this must have taken one man about eighty days and cost the modern equivalent of about £8,000.

In other words, hedge-laying was *always* expensive in the short term – but it brought huge rewards, for decades and centuries ahead.

Hedge-Breaking Times

A fair amount of the medieval landscape remained the unhedged open-field system, however, especially in the East Midlands and the North, so today's hedgerows there are often younger. In Huntingdonshire, a centre of open strip-field farming, hedges are mostly early nineteenth century in origin, whereas in East Kent and Sussex, they tend to be very ancient due to much earlier enclosure as far back as Anglo-Saxon times.[10] One famous Huntingdonshire exception, though, is Judith's Hedge at Monks Wood research centre, which was planted by the countess Judith of Lens, a niece of William the Conqueror, who received the land in 1075 when her husband was executed for high treason. (Although, given that she appears to have played a rather unsavoury role in the death of her husband, Waltheof, the Earl of Northumbria, I'm not sure she *deserves* to have a hedge named after her.)

But both open field and hedged landscapes were once interspersed with many woodlands, heaths, 'waste' and commons available to all. It was the enclosure of these commons by wealthy landowners with fences and hedges, from the Elizabethan period onwards, that gave rise to a new word in the English language: 'hedge-breaker', as the rural poor, deprived of their ancestral grazing rights on the open commons, set about tearing down the hated new barriers in protest.

Thus, Kett's Rebellion in Norfolk, in 1549, led by the fiery Robert Kett. 'The common pastures left by our predecessors for our relief and our children are taken away. The lands which in the memory of our fathers were common, those are ditched and hedged in and made several; the pastures are enclosed, and we shut out.'[11] Kett was eventually hanged from the walls of Norwich Castle.

A gentler bye-law from Saffron Walden in Suffolk, from 1561, decrees that: 'Every hedge-breaker taken in breaching of hedges and carrying off suchlike wood shall pay ... 16d. and 3 hours punishment in the stocks ... 4d to the lord, to the owners of such wood 4d, to the bailiff 4d., and to him that shall take an offender 4d.'[12] But back then, 16 pence was a lot of money to a poor family.

During the Civil War, the term 'Leveller' gained currency, often assumed to mean a proto-egalitarian bent on levelling social distinction, but actually it denoted those who wished to level the hedges and regain common land, while their fellow Diggers dug and planted where they chose, not recognising private property rights, their leader Gerrard Winstanley talking of 'The earth (which was made to be a common treasury for all, both beasts and men)'.[13]

Enclosures also put many people out of business, including shepherds, cowherds and swineherds, whose job was to keep an eye on free-roaming beasts on common land, but who were no longer needed once the animals were confined to hedged fields.

Nevertheless, some common lands still exist across England, including the New Forest, with grazing rights for residents. And we shouldn't imagine that *all* boundaries and hedges were disliked on principle – only those that deprived people of their common rights.

Over time, our hedgerows began to reveal the wonderfully varied regionalism of England as well, based ultimately on its fantastically heterogeneous geology. 'The UK and Ireland feature some of the most diverse and beautiful geology in the world, spanning most of geological time, from the oldest Pre-Cambrian rocks to the youngest Quaternary sediments,' explains the Geological Society on its website; while in comparison, enormous swathes of Siberia, say, are… well, geologically *uneventful.*

Geology determines soil type, of course, as well as local building materials and styles, from the granite farmhouses of Cumbria to the flint and brick of Norfolk, from the stunning red sandstone buildings of Herefordshire to the cob and thatch cottages of Devon.

Less often noticed, though, is how this geology also affects the species that inhabit our hedgerows. The regional differences can be striking. In Staffordshire, holly hedges are common on the loamy soil. Midland hedges tend to be mainly hawthorn, ideal for immensely strong 'bullock hedges' because the Midlands were traditional cattle country. In the wetter west and Wales, blackthorn flourishes – good for snagging and restraining sheep – and even more so in Ireland, where this redoubtable shrub produced the wood for the Irishman's time-honoured shillelagh or blackthorn stick, its incredibly hard root ball forming a useful natural club for fighting one's enemies – or possibly the English.

Dykes in Norfolk were sometimes known as wet hedges, while what we could call an earth bank in Cornwall was called a hedge, often topped by gorse or tamarisk. Breckland on the Suffolk–Norfolk border, 'England's desert' as it has been called, boasts impressive shelter belts of pine as well as some unique lilac hedges; the Somerset Levels have hedges of osier; and the hop gardens and orchards of Kent and Worcestershire have always favoured large-leaved trees like beech, poplar and

35

alder, growing up to thirteen feet high, to protect their valuable crops from the wind. And then there are the beautiful hedges of sea buckthorn around the East Lothian coast; a hedgerow shrub, incidentally, that would grow pretty much anywhere, if planted, and produce bumpers crops of gorgeous and highly nutritious orange berries every year – as they do on our land in Wiltshire.

Varieties of hedge around the country even have their own names, such as the Leicestershire bullfinch, a bull-proof hedge of immense strength, both fun and foolhardy to try and jump when on horseback in pursuit of a fox. Then there's the Glamorgan flying hedge, the East Anglian arable-land coppice hedge, the Wealden shaws of Sussex, the Scottish backit dyke...

Like the thatched cottages of Devon, the slate roofs of North Wales, the half-timbering of Suffolk and the granite farmhouses of Cumberland, these different styles of hedgerow across the countryside are testimony to our ancient diversity and richness, increasingly erased by boring modernity. Across Europe, meanwhile, one ecologist has identified a staggering 135 different management styles for hedges.[14]

While we already had countless thousands of miles of centuries-old hedgerow by the eighteenth century, the great period of enclosures in Georgian times vastly increased them again. It's been reckoned that, in all, they enclosed a staggering seven million more acres of land, creating thousands more miles of hedgerows loved by wildlife, but hated by the rural poor, who again lost more of their valuable common lands and rights.

'Plots of land allotted by virtue of this act shall be inclosed and fenced round with ditches and quickset hedges with proper posts, rails and other guard fences,' explained one eighteenth-century gentleman.[15] Quickset means hawthorn, also called quickthorn, may or even hagthorn. A hawfinch, incidentally, means a hawthorn finch, first named in 1676 by the Warwickshire naturalist Francis Willughby. (He also wrote a treatise called 'On the Rebounding of Tennis Balls', which sounds simply fascinating.) One can't help noticing that the name hawfinch was thus bestowed just about the time enclosure was getting going. Did the tremendous increase in hawthorns being planted as hedgerows across our countryside result in a boom in hawfinch numbers?

By around 1820, we had around half a million miles of hedgerow, or about 700,000 kilometres, as compared to today's diminished 400,000 kilometres, replicating quite by accident one of the greatest edge habitats on earth. Perhaps only one other such edge habitat or liminal zone, the seashore, is ecologically as important.

Our ancestors, observant of nature as ever, understood the tremendous power of these half-and-half zones, neither quite sea nor land, neither quite forest nor grassland. In pagan rituals, for instance, if you really wanted to cast a spell of great effectiveness, you did so at a crossroads (neither one road nor another) and at midnight (neither one day or another).

The golden age of hedges and hedge-planting – broadly 1700–1850 – meant that 'A square mile of arable land in East Anglia may have contained about ten to fifteen miles of hedge and the South-west, twenty-five to thirty miles.'[16] And so many miles of hedge also meant it was the age of skilled hedge-laying. One can still see the different, highly craftsmanlike styles illustrated in old almanacs, for instance in R.W. Dickson's *The Farmer's Companion* of 1813. Little wood was wasted. 'Even the tiny twigs, known in Devon as "spray wood", were not thrown away: they were used to make charcoal and to fire bakers' ovens.'[17]

It was probably at this time that many folkloric practices developed around hedges, although they may have their roots much earlier. At Great Gransden in Huntingdonshire, says Oliver Rackham, 'when "cessioning" the bounds – that is patrolling, inspecting and asserting their rights over the land – the villagers used to dig a hole at a certain spot and hold the Vicar's head in it.'[18] (Presumably the head was still attached to the vicar's body.) Definitely older, so old indeed that no one knows quite what it signifies any more, but we carry on doing it anyway because we're *British*, is the ceremony of the Penny Hedge, also called the Horngarth, at the Upper Harbour in Whitby. This involves the construction of a short length of hedge strong enough to withstand three tides, on the eve of Ascension Day, while everyone shouts: 'Out on ye! Out on ye!' No doubt it all makes sense if you're there.

Hedges were also carried abroad by the British, to New England, Tasmania and New Zealand, all of which have some excellent examples of hedgerows, although beyond the scope of this book. John Dover finds them even more widespread, in virtually every continent, and no doubt of great ecological value every time.[19]

The Decline of Our Hedgerows Today – and How to Fight Back

Hedgerows did continue to increase towards modern times. Between 1750 and 1850, estimates Richard Muir, hedges were being planted at a rate of two thousand miles a year, creating the now familiar dense patchwork landscape of Britain's countryside.

But since the 1950s, and accelerating into the 1960s and 1970s, our landscape has been flattened out by the catastrophic hedgerow losses that we have now become so familiar with.

Entomologist Stephen Jones writes that the peak of hedgerow destruction in the 1960s saw a rate of hedgerow loss of sixteen thousand kilometres a year.[20]

This has also been true across Europe where modern farming dominates, with John Dover mentioning a 'land consolidation programme' in Brittany in 1982, which led to a 50 per cent reduction in hedgerow.[21]

Even more barmy, the EU, as George Monbiot reminds us, once fined the Northern Ireland government for 'giving subsidy money to farms whose traditional hedgerows are too wide'.[22] But then this is the same EU that has also punished or threatened farmers who allow the lovely yellow flag iris to grow in their fields – 'encroaching vegetation', they classify it – as well as gorse and dog rose. They even fined farmers who allow more than fifty trees per hectare of agricultural land, at which point, according to the visionary geniuses of Brussels, the field instantaneously becomes woodland and therefore can no longer be farmed. (Not that our lot in Westminster are necessarily any better, by the way.)

The loss of ancient hedgerows is particularly lamentable and seems to be accelerating, if anything. Often it is done by 'accident' or 'oversight' by developers. For instance, as recently as 2022, Barratt Homes destroyed an ancient hedge just outside Bridport, Dorset, to make way for 760 new homes. Barratt Homes and Vistry Group were forced to publicly apologise following the incident, which they put down to a 'communication failure', adding they will be replacing the removed elements 'as soon as possible'.[23]

Said one angry local, 'The developers' claim that they will put this destruction right is laughable.'

Quite so. You cannot just 'replace' an ancient hedge by planting a new one.

Bellway Housing has destroyed an ancient hedge at Hilperton, Wiltshire, while Castle Green Homes destroyed ancient hedgerows at Prestatyn, North Wales.[24] And so on.

Hedgerow removal on farmland is now regulated and often forbidden but it still continues, while housing developers, it seems, have carte blanche. Hedges need far more resolute legal protection than they currently enjoy, and developers who break the rules should be prosecuted every time. But with the government's own hated HS2 rail development destroying or damaging literally dozens of ancient woodlands, what chance is there for hedgerows?

The CPRE provided an invaluable analysis of illegal hedge removal and its frequency of prosecution by local councils for the year 2010, and it makes grim reading. In the South West, for instance, 3.67 kilometres of precious, ancient and legally protected hedgerow were destroyed in 2010. Just 0.5 kilometres were brought to successful or 'pending' prosecution. In the East Midlands, 11.01 kilometres were destroyed – and the successful prosecutions? Zero.

As with so many other environmental crimes, if you destroy an ancient hedgerow, you'll probably get away with it. Is it too cynical to suggest that developers and house builders have noticed this?

Today, the greatest threat to our ancient hedgerows isn't actually their removal, but something more insidious and less obvious: their gradual deterioration. Unless a hedge, at least to some extent, resembles a dense hawthorn or blackthorn thicket, it will be of far less ecological value. And if it has deteriorated into little more than a blunt, dwarfish line of scarred and wind-scoured stumps, devoid of any accompanying grassy verge or ditch, it's of less value still.

The sad truth is, while our remaining 400,000 kilometres or so of hedgerow may sound pretty impressive, if almost half of these are just such lines of stumps, then the situation is worse than we realised.

It's all too easy for a farmer or landowner – indifferent or overworked as they may be – to let this relict, much-flailed hedge die off altogether. On arable land it may be seen as an obstacle, while on pastureland, livestock can be separated and managed instead with that ugly modern invention: the barbed-wire fence. First devised in the 1860s, in either France or the USA, depending on your sources, it made possible the rapid westward expansion of farming across the Great Plains and was a key factor in the decline of the American bison.

Today, it is still having a ruinous effect, by offering a cheap and easy alternative to hedges. Barbed wire, by displacing our traditional hedgerows, is far worse news for wildlife than is often realised. And its supposed 'cheapness' is, as with so many other things, a short-term illusion. Here are some ways a 'cheap' wire fence differs from a hedge:

It provides no windbreak, prevents no soil erosion, stores no carbon dioxide, emits no clean oxygen, absorbs no surplus rainwater and therefore does not mitigate flooding. It provides no wild foods, berries, nuts, wild greens or herbs, hedgerow medicines or fodder for animals, and it offers no shelter or habitat to birds and mammals, beneficial pollinators and insect predators.

It does, on the other hand, require fence posts, which will need to be renewed every ten to twenty years or so, and are treated against rot. Until 2004, our fence posts were treated with chromated copper arsenate, an arsenic-based carcinogen that can cause a whole range of cancers, cardiovascular disease and more. Chromated copper arsenate may now be banned, but there's still plenty of it out in our environment.

Nowadays, fence posts can still be treated with old-school creosote (which also causes skin cancer, lung cancer, nausea, vomiting, etc.), along with such ingenious new delights as didecyldimethylammonium chloride. Do we even *need* to do extensive tests to know whether this is safe or not?

Nope. Given that we share 70 per cent of our DNA with slugs and 60 per cent with bananas then, as with every other human-made chemical, pesticide and insecticide, the same rule applies: if it's designed to be toxic to other organisms, it'll be toxic to us.

Meanwhile, amazingly, a living conservation-laid hedge manages to flourish and grow and strengthen and not rot in the ground *all on its own.*

And then of course there's the barbed wire itself. We don't produce enough steel in the UK, so we import around half a million tonnes from China every year, along with another six million tonnes from other countries. The environmental costs of this are, needless to say, enormous. And when the barbed wire eventually rusts and falls apart, it might be recycled – but only in an electric arc furnace running at 3,000–3,500°C. And where does this energy come from?

You can guess.

So, while the superficially 'cheap' wire fence actually requires a massive CO_2-spouting global-industrial infrastructure to exist at all, a living hedge can be cut and laid by one person with an axe and saw over the course of the winter, and last indefinitely. As for new hedgerows, I've planted some on our land simply by digging up small, overshadowed seedling trees from our woodland in winter and sticking them in the field where required: total supply-chain distance about one hundred yards, rather than from China. I'm sure this is how many ancient hedgerows were created too. John Evelyn, in his wonderful *Sylva, Or Discourse of Forest-Trees* (1664), says specifically that this is how he has planted his holly hedges: 'I take thousands of them four inches long out of the Woods, (amongst the fall'n leaves whereof, they sow themselves).'

As with so many areas of modern life, we seem simply incapable of correctly costing things – especially over the longer term.

As this book aims to demonstrate from many different angles, the benefits of our immense network of veteran hedges, especially if conservation laid, in a country whose wildlife and arguably entire ecosystem stability is in serious trouble, are almost incalculable.

The flat arable lands of East Anglia used to have splendid hedgerows but have suffered the worst losses since the war, and especially since the 1960s. Hedgerow removal was actually subsidised first by the UK government and then, after 1973, far more zealously by the EU and the catastrophic Common Agricultural Policy.

John Dover records a personal anecdote of a large estate in Cambridgeshire, where a wealthy but not terribly bright new landowner, in the interests of 'increasing farming efficiency' and the numbers of game birds for his shooting business, grubbed out many of the hedgerows as useless obstacles, sacked staff, sold off tied cottages, and contracted out ploughing, spraying and harvesting to cheaper machine operators with no interest in or love of the land. When he discovered that there were hardly any game birds left to shoot, because they'd lost their much-needed cover, the landowner had to start replanting the whole lot from scratch.

This is a particularly dumb example, but only an extreme version of what has happened nationwide, where we are now having to replant lost hedges at great expense.

And while hedgerow removal may give more space to larger machines in ever more prairie-like fields, it frees up very little land. 'In order to produce one extra acre of land for cultivation it is necessary to remove 2,420 yards of a six-foot-wide hedge – or to produce an extra hectare, 5,000 metres of a two-metre-wide hedge.'[25] No doubt yields would actually decline as a result of increased wind speeds, soil erosion and so on.

Another hidden loss, of a part of the hedgerow that is overlooked but is most certainly a key constituent, is the grassy margin alongside. Immediately next to shelter and nesting sites, it should be a rich feeding ground. And nationwide it adds up to a vast area. Sue Clifford and Angela King supply this startling calculation of the area of roadside verges in the UK alone: 'The green estate alongside Britain's roads covers some 523,000 acres – the size of Surrey.'[26]

Where landowners and councils manage margins and verges with some intelligence – roughly speaking, confined to a late summer 'hay cut' only – they may contain rare and very localised flora: Devonshire lanes for bastard balm; Gloucestershire's Severn Vale for lady's smock; the Lizard in Cornwall for wild madder, navelwort and three-cornered leek. And like much of our wildlife, they connect to our past as well. 'Perennial flax is a good sign of a Roman road on chalk or limestone verges; it grows along Ermine Street and the A11, just north of Stump Cross, a "Romanised" stretch of the Icknield Way marking the Essex/Cambridgeshire boundary.'[27]

From the Palaeolithic to the present, as this brisk gallop through our history shows, our native hedgerows have had a long presence in our landscape, especially when taken together with their forebears, the thorny thickets of the wood pasture before them. This is the key to their value, and the plethora of wildlife benefits that they provide, as we shall explore in subsequent chapters.

Additional Note:
Dating a Particular Hedge

This can be great fun – but it *is* a bit of minefield, and you may just have to conclude that a hedge is 'quite old'. This was our final conclusion with the hedge at Underhill: we could be sure of a certain amount, but, ultimately, I would say it is 'centuries old' and leave it at that.

One sure way of knowing is finding a current hedgerow clearly marked on the extraordinarily detailed tithe maps that were created in the early Victorian period, from 1837 to 1856, covering the whole of England and Wales. They were drawn up with the aim of bringing the practice of obligatory 'tithing', that is, paying a portion of all agricultural produce to the Church, to an end, and are now available to view online at www.thegeneal-ogist.co.uk/tithe.

The Underhill hedge, happily, is very evident on the East Knoyle tithe map, so we can be certain that it's approaching 200 years old. There is, as yet, no official classification for an 'ancient hedge' as there is for as woodland: ancient woodland, by definition, must date from at least 1600 in England and Wales, and 1750 in Scotland. But it is generally accepted that any hedgerow already existing before the eighteenth-century enclosures got going may be considered as ancient.

There are other remarkable old maps too. Britain is unique in having been mapped directly, at hedgerow and field level, in the Ordnance Survey's majestic 1853–1893 series, where every mile of our country occupies six inches of mapping. This will surely never be repeated and is a gift to the amateur landscape detective wanting to trace hedgerow history. In the Second World War, the Luftwaffe photographed much of southern Britain in detail from the air, another invaluable record – although, to be fair, they weren't exactly doing this out of concern for conservation. Yet the photos revealed, says Oliver Rackham, 'what was still, in many places, a medieval landscape', with fields and boundaries, lanes and churches unchanged for six or seven hundred years.[28]

Back beyond these maps, though, certainty about the Underhill or any other hedge gets trickier, although the landscape around is pretty old, with human habitation evident from the Mesolithic, plenty of Bronze Age and Iron Age barrows in the vicinity, and a church with a chancel that dates from pre-Norman times. It would be safe to guess that most hedges here are at least Anglo-Saxon.

Then there's the classic 'Hooper's Rule', which was first proposed by undoubted hedges expert Dr Max Hooper back in 1965:

Age of hedge = number of woody plant species in a 30-yard section x 110.

So, if a 30-yard section of hedge contained hawthorn, blackthorn, elm, spindle and crab apple, say, you could reckon it was 550 years old.

Now, Dr Hooper was a bona fide hedges hero. It was he who first pointed out back in the 1960s that Britain was losing an appalling ten thousand miles of hedgerow every year, by studying old black-and-white photos taken of the English countryside by RAF reconnaissance pilots during the Second World War. He also warned that his rule was very much a rough guide. But I'm sorry to say that, along with many others nowadays, I find it pretty unconvincing.

For example, there's a hedge just down at the bottom of our field, around the orchard, which contains twelve different woody species, including such rarities as wild pear and wild privet. This would suggest that it's an impressive 1,300 years old or so, first springing to life about the same time Offa of Mercia was digging his famous dyke. But… this hedge is precisely 11 years old, the same as the young orchard it encloses. I know because I planted it myself.

A certain common sense and an ability to read a landscape is often as good a guide as any. Is the hedge thick and dense? Is it in a seemingly 'old', unchanged landscape? Does it include massive, mossy earthen banks? And does it contain large standalone trees, especially oaks, which could be roughly dated? All of these features are present in the Underhill hedge.

And remember, there is a broadly reliable rule of thumb for dating oak trees.

If it would take two people to hug it completely, it's about 150 years old.

Four people: it's about 400 years old.

And six people: 1,000 years old.

(I love this rule. Apart from anything else, it's a scientifically justified reason to go around hugging trees.)

Further clues might be had from studying the names of the individual fields: again, these can be often be found on old maps, as well as the tithe maps. Certain field names tell you that the fields have been cleared from woodland or wasteland: they include Stocks, Stocking, Stubbing, Riddens, Riddings, Sart, Stubbs and Rode. The hedges around them are likely to be very old.

There are further pointers to the age of a hedge from the flowering plants that it supports, for instance the presence of ancient woodland trees such as small-leaved lime or wild service tree. But, in general, if you're lucky enough to encounter, or even own, a hedge that is tall, dense, thickety, diverse with different woody species and *looks* like it has been there a long time – then it probably has.

CHAPTER TWO

The Green and Brimming Bank

Hedges and Plants

I love these overblown and unruly hedges, and
of all the jumble that clings about them I like the
dog-roses best. Oh, those June roses in Suffolk ... !

Julian Tennyson, *Suffolk Scene*

A staggering 1,000 plant species or more have been recorded in British hedgerows over the years, nearly a third of the total of some 3,500 or so species native to the UK. 'If a wide verge is included, then there is little limit to the number of species that may be found', while one rarity, the Plymouth pear, is found *only* in our hedgerows.[1]

Some of our loveliest wildflowers – spiked rampion, spreading bellflower and hedge bindweed – are rarely found except in hedgerows either, reminding us what a rich refugium of wild nature they are in an otherwise brutally denatured countryside.

To discover just what plants comprised the hedge at Underhill, we asked a biologist friend of ours, Jenny Bennett, to do a complete survey for us. And while it's true that we might have managed to identify hawthorn and hazel for ourselves, there's no way we could have identified everything here.

On a fine spring morning in early May, this was the list that Jenny came up with, in this one, thirty-metre, conservation-laid hedge:

Trees and Shrubs

Ash (*Fraxinus excelsior*)
Blackthorn (*Prunus spinosa*)
Bramble (*Rubus fruticosus* agg.)
Dog rose (*Rosa* spp.)
Field rose (*Rosa arvensis*)
Hawthorn (*Crataegus monogyna*)
Hazel (*Corylus avellana*)
Oak (*Quercus robur*)

Ground Flora

(F) *denotes frequent.*
* *denotes ancient woodland indicator.*

Barren strawberry (*Potentilla sterilis*)
Bluebell (*Hyacinthoides non-scripta*)*
Bush vetch (*Vicia sepium*)
Celandine (*Ficaria verna*)
Cleavers or goosegrass (*Galium aparine*)
Cock's-foot (*Dactylis glomerata*)
Common polypody (*Polypodium vulgare*)
Cow parsley (*Anthriscus sylvestris*)
Dandelion (*Taraxacum* agg.)
Dog's mercury (*Mercurialis perennis*) (F)
Early purple orchid (*Orchis mascula*)*
Germander speedwell (*Veronica chamaedrys*)

Ground ivy (*Glechoma hederacea*)
Hart's-tongue fern (*Asplenium* or *Phyllitis scolopendrium*)
Herb bennet (*Geum urbanum*)
Ivy (*Hedera helix*)
Lords-and-ladies (*Arum maculatum*)
Meadow buttercup (*Ranunculus acris*)
Meadow foxtail (*Alopecurus pratensis*)
Meadow vetchling (*Lathyrus pratensis*)
Nettle (*Urtica dioica*) (F)
Pendulous sedge (*Carex pendula*)
Pignut (*Conopodium majus*)*
Ramsons (*Allium ursinum*)*
Soft rush (*Juncus effusus*)
Soft shield fern (*Polystichum setiferum*)*
Wood dock (*Rumex sanguineus*)
Wood false-brome (*Brachypodium sylvaticum*)
Yorkshire fog (*Holcus lanatus*)

There are no fewer than five ancient woodland indicators here, which was a terrific sign; we were thrilled, and pleasantly surprised, to realise there was an orchid colony in the hedgerow. There was also a mighty mature oak tree, so again this would suggest that the Underhill hedge is very old and ecologically valuable – but still perfectly happy to be laid and laid again in the traditional manner, as the need arises, springing to magical new life each time.

A survey back in 1957 found the twenty most frequent species in comparable hedgerows across Wiltshire to be these: ivy, cleavers, hawthorn, bramble, wild rose, stinging nettle, ground ivy, false oat grass, ash, hogweed, cuckoo pint, blackthorn, elder, hazel, hedge woundwort, cow parsley, herb robert, field maple, hedge garlic and wild privet.[2]

Wild privet is a characteristic shrub of chalky soils, so it loves the wide Wiltshire downs, but we would rarely expect to see it in our own corner of the county, which happens to be more loam and greensand. Otherwise, you can see there's considerable overlap in the two findings, and a great sense of continuity – as there was for generations of people and how they viewed, cared for and used their precious hedgerows.

Hedge Trees

Hazels are one of the commonest trees in our hedgerows, and the most satisfying and easiest for traditional hedge-laying, with multiple shoots and nice springy, forgiving timber. They can also be coppiced very easily, of course, responding just as they would have done to hard browsing by our vanished megafauna thousands of years ago. It's always a joy to work with hazel.

But blackthorn and hawthorn are the two absolute staples of any stock fence, being densely thorny and hard for even the most determined sheep or cow to bulldoze through. The two thorns also have multiple other uses, however, and of course are partly edible to us as well as to livestock, especially the berries. Hawthorn twigs were formerly favoured for rake tines, and the blackthorn, to this day, makes the best of all walking sticks.

Spindle, with its highly distinctive pink-and-orange berries in autumn, was much valued in hedgerows, being carefully harvested to make spindles for spinning wool; elder was used for clothes pegs; dogwood's smooth straight shoots were used as skewers for roasting meat, while a tea made from dogwood bark is a powerful painkiller – as is willow bark. Dogwood berries were also collected and pressed to yield an oil that was made into soap and even oil lamps. Apparently, it burns very cleanly and well.

The qualities of various woods for the carpenter were also known in great detail. Elm is very resistant to rot and splitting, so its disappearance from our landscape has been a great loss not just to wildlife but also to rural craftsmen. Its timber was used for coffins, weather boards and water pipes, 'whole trunks being bored out and driven together to make mains pipes'.[3] It was also invaluable for village pumps, the hubs of cart wheels, and indeed anything where the timber needed to remain strong while frequently encountering water.

In every case, we should picture our wise and frugal forebears carefully *harvesting* the timber from living hedgerows and then letting them sprout again, rather than cutting them down. Pollarding and coppicing mean, respectively, either cutting off all the shoots and branches from a tree at about six to eight feet in height – as high as someone with a saw can reach, in other words – leaving just the main trunk, or else cutting the entire tree down to ground level but leaving the stump or stool in the ground, with complete root system. Far from killing the tree, either of these techniques encourages it to shoot again with tremendous vigour, thanks to that widespread root system. Of course, there was also clear-felling and large-scale forest clearance in the past to create agricultural fields, but once these were established, the remaining copses and hedgerows, so easily accessible to the woodcutter, were far too valuable to destroy entirely. And removing a tree stump from the ground, along with its tenacious root system, is colossally hard work. Nowadays, they use powerful machines like diggers or stump grinders, but in the old days it would have taken whole teams of men and draft horses. Why bother, if you don't really need to?

Happily, trees aren't easy to destroy. Many will sprout back again from cuts and stumps – as in the pollarding and coppicing just mentioned – and / or send out vigorous suckers underground to produce identical copies of themselves. It's remarkable to realise that, thanks to such suckering in certain species, some hedgerows may be made up of the same one family of trees, or even clones of a single parent. Pollard, Hooper and Moore, in their book *Hedges* in the revered New Naturalist series, suggest that if ever you see an entire row of elms along a field boundary, they're probably all suckers from the same tree; and, similarly, many blackthorn hedgerows might well be from a single or only a handful of originals, all closely related siblings and family members.

This is also true of oak trees in parkland and wood pasture. If you identify a very large 'mother oak' some five or six centuries old, along with several 200-year-olds nearby, they are highly likely to be, quite literally, her offspring: planted by jays, still living close to mum and still nourished by her through underground mycelial networks. A single teaspoon of good soil can contain literally *miles* of these fungal filaments – and trees communicate through them too, in electrical pulses that travel at the very sedate pace of one third of an inch per second.[4]

It sounds anthropomorphic but it's true. Although as Peter Wohlleben points out in his wonderful *The Hidden Life of Trees*, many small oak seedlings, growing up in the shade of their gigantic mother, can only really get going in life once she starts to die off or comes crashing down in a storm, letting the light in. This no bad thing: growing very slowly in their formative years, the young oaks will have developed very dense timber, and often be far more disease resistant than young sprouts that grow up too quickly in full sunlight.

Hedgerows sucker rapidly into adjacent pastureland if allowed to. Judith's Hedge, for instance – the kind of overgrown and fast-spreading hedge that tends to delight a naturalist rather than a farmer – Pollard notes is full of suckering species, especially dogwood and blackthorn, whose suckers have invaded the adjacent grassland. This naturally creates sheltered bays and recesses whatever the wind might be doing, which is a wonderful thing for wildlife. The best hedge will be winding and scalloped, not dead straight, and this can be achieved most easily simply by allowing room for suckers to spread and create a natural range of bays and inlets, similar to what ecologists called 'nodes'; that is, hedgerow intersections and corners, which are richer habitats again than straight hedges, with nesting bird numbers nearly double there than the hedgerow average.

Elms' energetic suckering means they're great in a newly planted hedge, and if not permitted to grow above a certain height they shouldn't succumb to Dutch elm disease, which is why you can still find plenty of elm in hedgerows.

Another more unusual sucker in British soil, I've found, and I'm sure in its native eastern North America, is the chokecherry or chokeberry, *Aronia melanocarpa*, which I have planted on our land for its incredibly nutritious fruit and which has sprouted offshoots with great enthusiasm. Of course, there are arguments for only planting native species, but perhaps these can be overdone – by the same token, one shouldn't try to grow figs or grapes in Britain either – and it might be more reasonable to aim for 90 per cent natives while allowing a few other exotic foreigners into the mix. Though one truly, almost comically useless foreign introduction is the plane tree, first brought over here in 1520, and still apparently supporting *no species at all*.[5]

Pollard, Hooper and Moore's original findings were confirmed in the important paper 'The Value of Different Tree and Shrub Species to Wildlife' by Keith Alexander, Jill Butler and Ted Green, which appeared in *British Wildlife* in

2006, and makes fascinating reading.[6] And yes, the plane tree remains pretty useless: the authors rate trees with a user-friendly one- to five-star rating on various criteria, five stars being the best. And the London plane actually scores *zero* stars in several key categories, including foliage inverts, wood decay inverts, blossom for pollen and nectar, and fruits and seeds.

However, not all non-native species fare as badly. We already know that the hated non-native conifers of our gloomy plantations do at least support populations of firecrest and crossbills. And as the paper reveals, Norway spruce and European larch both score five stars for mycorrhizal fungi, and four stars for fruits and seeds. So, for the latter, very important class, they are actually more important than natives like elm or hazel.

The list might offer you a guide if you're planting a new hedgerow: a windbreak of Norway spruce wouldn't be as bad as some might think. And I certainly feel less guilty about our fifteen or so chokecherry bushes, which produce abundant dark berries that are great with stewed apple, contain five times more antioxidants than the blueberry, more than any other northern temperate fruit, and are adored by the birds in autumn – as well as deer and, inevitably, our cows.

The most redoubtable sucker of all though, in my experience, is still the good old blackthorn, which will march out into a field in great numbers like an invading army and may actually have to be restrained at some point for balance. Still, like hawthorn, blackthorn is a wildlife essential in any hedge: of moths alone, 'over a hundred feed on hawthorn and nearly as many on blackthorn', and the latter is also crucial to the threatened brown hairstreak butterfly, if not flailed relentlessly every year, since it lays its eggs on young blackthorn shoots.[7]

Queen of the Trees

Queen of May, Queen of the Hedgerow, is the hawthorn, named after the hedge itself: the hagthorn, hegthorn, hedge-thorn – it's all the same thing. In Ireland, it has the lovely Gaelic name *sceach geal*, which means 'bright thorn' or 'moon thorn', either of which is wonderfully evocative of its great swathes of glowing white blossom in early summer.

It will provide a home for at least three hundred species of insect, including the one hundred or so moth species mentioned above, with such fabulous names as the orchard ermine, the pear leaf blister, the rhomboid tortrix, light emerald, lackey, vapourer, fruitlet-mining tortrix, small eggar and the lappet moth. Hawthorns can live for up to five hundred years, but often start their life as seeds passing through a bird's digestive system, which can speed up germination by a year compared to uneaten haws.

Of hawthorn and blackthorn together, says Benedict Macdonald, 'Their presence in a landscape is as important, in their first fifty years of life, as oak is in its last few centuries.'[8]

Incidentally, while it is now well known that oaks support the greatest variety of invertebrate species and others, the rest of the list of trees, in descending order, is a fascinating and important one.

Second place goes to willow, then birch, hawthorn, blackthorn, poplar, apple, pine, alder, elm and hazel. And all of these species love as much light and sunshine as they can get; in other words, they are adapted to wood pastures and open grasslands, not damp dark forests.

While we all view planting oaks as the gold standard for regenerative farming and eco-friendly tree planting, it only increases the importance of thorny traditional hedgerows to know that hawthorn and blackthorn come fourth and fifth on the list of importance, with hazel following at number eleven.

And willows may still be overlooked by would-be wildlife-friendly tree planters and rewilders, supporting nearly as many insects as do oaks. Remembering that the UK in its natural state would have been as much as 20 per cent fen, marsh and wetland of one sort or another, and given the sheer numbers of willows accompanying these watery domains, it's no surprise that a lot of our native wildlife would also be very adapted to willow.

Again, we're reminded that preserving our established, hawthorn-and-blackthorn hedges, cutting and laying intelligently rather than just flailing, and planting more, will have far more *immediate* benefit for wildlife and our countryside than starting any number of young oak plantations. Properly laying and then leaving a hawthorn hedge alone, instead of flailing it, leads to a flower and fruit increase of a whopping 75 per cent the very next year. Some quick fun with numbers: hawthorns are found in over 90 per cent of all our hedgerows, so we might estimate that 10 per cent of all the hedgerows across England are made up of hawthorn. Of our 400,000 kilometres of remaining hedgerow, that's about 40,000 kilometres of hawthorn.[9]

By simply not flailing for a year, or by flailing on only alternate years, we thus increase the hawthorn flowering, nectar and fruit by 75 per cent over 40,000 kilometres of hedgerows. This would have a huge effect on insect populations and everything else. Ditto blackthorn and bramble: blackberries actually provide the highest energy content of all berries in autumn, thanks to the sugars (you can tell by the taste), and berry numbers again will increase 2.6 times over by 'incremental cutting'; that is, cutting each year but allowing for growth, so that the hedge, though trimmed, gradually gets higher and wider with time, instead of simply being knocked back every winter to the same old square dimensions, as happens with flailing. It could reasonably be argued that brambles are the single most important species in the hedgerows after oaks, as well as the most despised, the most often slashed and removed. Admittedly they can rather take over a hedge if allowed to, but in terms of biodiversity they are uniquely precious: all four thousand or so microspecies of them. Even their 'cut stems are a major nesting site for various small solitary bees', notes Buglife.

With mere browsing by livestock, however, hawthorn grows back stronger, thanks in part – a very recent discovery, this – to the vitamin B_1, or thiamine, in the saliva of cows, deer, etc., which stimulates plant growth. But the tree will also put out ever longer thorns, to dissuade the beasts from eating more than they should, thus becoming what Benedict Macdonald calls 'scrub fortresses': many a bird's dream home when it comes to safe nesting.

But if the farmer wanted to keep over-eager animals back from the hedge for a season or two, then he could always make a hurdle or pole fence from the hedge itself (see The Wonder of Pollards on page 61).

From yellowhammers (now seriously threatened – down 61 per cent since 1967) to wood mice to slow worms, which are really lizards, numerous species will take up residence in the thick, thorny protection of a hawthorn hedge, although they prefer one that is not too high. If you're really lucky, and you live in south-east or south-central England, and the surrounding countryside is amenable (a big if), you may even attract a nightingale to your hawthorn hedge-thicket, their favoured place of all, especially if heavily overgrown with sprawls and swags of bramble and wild rose.

In *Wiltshire Village* by Heather and Robin Tanner, first published in 1939, the authors make the following observation: 'For several years in succession a pair of nightingales came to White Stocking Lane, but one winter the hedges were ruthlessly trimmed and the lane cleared. Perhaps the nightingales did not recognise the place the following spring; at any rate, they have not been since.'[10]

It can also protect us, if you believe the folklore. The hawthorn has one of the richest mythologies of all our native trees, starting with the very practical saying 'cast ne'er a clout til may be out', which means don't discard any layers of clothing – clout – until the hawthorn has blossomed in about mid-May! Our ancestors were well aware of the importance of not catching cold. But there's deeper matter in the mysterious connection of hawthorns with witches, or hags, from the Old English *hægtesse*, meaning witch but literally seeming to signify 'hedge-rider' or 'hawthorn-rider'. Similarly, Old Norse had *tunriða* and Old High German *zunritha*, both literally 'hedge-rider', used of witches and ghosts. It is all somewhat lost in the mists of time, but there was clearly some association.[11]

But whatever the etymology, the moral is plain: a hawthorn has magical powers for both good and evil, and you damage it at your peril. This seems a pretty good lesson for the current year to reflect on.

Hedgerow Landscapes as Scrublands

In combination in their thorny banks, hawthorn and blackthorn mimic our primordial scrublands.

The extensive stretches of oak-scattered wood pasture that ecologists now believe would have been the most common landscape throughout much of Britain's Holocene history since the last ice age, also included dense patches of thorn scrub, what Benedict Macdonald calls 'thorn-lands'. This is the 'natural' landscape of Britain and Europe without human intervention: not the dense closed-canopy forest, but something much more like the New Forest or Knepp. The reason primeval Europe would have looked like this was the actions of large herds of big, sometimes enormous, herbivores, grazers and browsers, constantly knocking back the trees and opening up more grasslands. It's very easy to forget what is no longer there, but if you picture roaming herds of bison, wild horses and one-and-a-half-tonne aurochs, you will get the picture.

Wide vistas of grassland, scattered oaks and thorn scrub. You can see how today's pasturelands of mixed grasses, thorny hedgerows and perhaps pollarded oaks, and/or the odd mature oak out in the field, are a very good yet unplanned approximation of this incredibly rich ecosystem, to which the great majority of our native species are adapted. Much more so, for instance, than dense conifer forests or even dense deciduous woodlands: good for mushrooms, maybe, but often surprisingly quiet.

The Wonder of Pollards

Before flailing, one of the most ubiquitous and important forms of hedgerow management that accompanied hedge-laying was pollarding. A style now extremely rare, alas, but you can see how common pollarding was formerly with a simple walk down any country lane: mature pollards are absolutely everywhere once you start to notice them.

The commonest trees to see pollarded nowadays are willows, often looking magnificent when bare and twiggy beside winter rivers, for instance on the Somerset Levels, in the Lincolnshire and Cambridgeshire fens, and the Thames Valley. But it can be done successfully to every species from field maple to holly.

In the past, while the rest of the hedgerow was laid and relaid in perpetuity, over generations, the mature standard trees that grew every thirty yards or so along the hedge would be pollarded;

that is, cut at about head height or a little higher, out of the reach of browsing animals. With an established root system perhaps already hundreds of years old, the pollarded tree would soon regrow new branches from its 'head'. In fact, the word derives from 'poll', meaning head; 'pollarded' more or less means beheaded – although like certain ribbon worms, and unlike human beings, trees that lose their heads can grow a new one.

Oliver Rackham cites the case of a 187-acre farm in Suffolk with an astonishing 6,058 pollard trees in its hedgerows in 1742 – one of those fascinating details that he was so brilliant at unearthing, that can evoke an entire landscape. Britain's hedgerows were once extraordinarily rich in mature pollards.

You can derive some idea of the importance of the managed hedgerow tree from Rackham's assertion that the timber for Nelson's navy came not from ancient woodland – there wasn't nearly enough ancient woodland left by then for such a purpose. 'The output of timber-built ships between 1800 and 1860 was probably equal to that in all the rest of history put together. Much shipbuilding timber, especially in large sizes and special shapes, came from hedges and parks, not woodland.'[12]

A mature tree has always had to compete with the hedgerow alongside it, but otherwise it has much less competition and far more sunshine than in a dense woodland and will grow faster and straighter than any such tree. It's a great invention.

The great benefit was the cuttings, which were invaluable: almost tree-trunk sized limbs could be taken off the top of a regularly pollarded veteran every fifty years or so, with no lessening of growing power. In the days when more than three hundred trees might be needed to build a single farmhouse, and many more for a large tithe barn or a hall, to produce endless timber by pollarding was only common sense. There were so many pollards, especially oaks, in our ancient hedgerows as well as our wide wood pastures, that even today Epping Forest alone can boast some fifty thousand veteran pollard trees. England's historic forests were widely spaced and sunny – picture today's New Forest, not some dark, dense, serried forestry plantation – and oaks respond especially well to maximum sunlight. In the rich grasslands of our former wood pastures, scattered pollard oaks would have produced abundant acorn crops in response to all that sunlight, which in turn would have greatly fattened the pigs in autumn. And then, every ten to twenty years or so, the pollards were cropped from the head or boll, the mighty timbers taken away for building materials: a supply that would just keep coming ad infinitum, unlike modern materials like breeze blocks, steel girders and cement, among the least sustainable materials ever devised.

Much lighter poles could be harvested from quick-growing hazel and willow, and be used to make pea sticks, beanpoles and hop poles, straw-bale pins and swill bools. (Swill bools? 'Rims for oak swill baskets.'[13]) Then there were thatching spars, charcoal, wattle and daub, hurdle fencing, besom brooms, handles for garden tools, fence posts, basketwork and barrel hoops. Pollards in Branscombe, Devon, were 'formerly managed as firewood for the village bakery'.[14] Bread was baked in the village, using an endless supply of firewood from hedgerow pollards not one hundred yards away. Where does the fuel come from today to bake our bread? And at what cost exactly?

Last but not least, pollarded (or coppiced) trees might provide the necessary materials for that essential household item: a coracle; that small, rounded one-person boat still sometimes seen in Wales, which can be made, say Rebecca Oaks and Edward Mills, in their excellent guide to coppicing, from just twenty-eight hazel rods of two centimetres in diameter and some deerskins. Now there's a project.

Pollarding for building timber took foresight and patience, long-term planning and a high degree of woodworking skill. The wood produced could also be used for fencing, firewood, charcoal and animal fodder and faggots. Ash and elm were usually cropped on five- to ten-year cycles, with strong ash being especially favoured for spear hafts and as handles for tools. Oak was left for longer to provide larger diameter timbers for construction, while the rarer 'hornbeam pollards that surround London were on a short rotation to supply the bread ovens of the city'.[15]

Like the hedgerow generally, pollarded trees weren't created to promote wildlife – but they did.

Most obviously, instead of growing in a long, straight trunk, as trees do in neat plantations, a pollarded tree will have a much shorter trunk and then suddenly branch out into a kind of goblet shape of many branches, which will catch much more sunlight, produce more blossom, seed and fruit, and makes ideal nesting habitat for birds. And far from being artificial, a pollarded tree is actually much more natural than the long, straight tree we are familiar with. Most trees in ancient woodland would have appeared coppiced or pollarded even without human intervention, due to repeated browsing and nibbling by large herbivores. I have even observed the effect with damage done to young trees by grey squirrels. In determined response, they simply put out a whole lot more shoots the following season.

The cutting of a pollarded tree, at the top of the 'bolling' – the gnarly, bulbous shape that develops at the top of the main stem – also produces numerous tiny pits and dents that will tend to hold rainwater, lead to a certain amount of fungal decay, but nowhere near enough to kill the tree. These micro-puddles, high in the air, can provide valuable breeding and feeding sites for numerous insects.

Larger pockets in the wood, and more advanced decay, will lead to cavities big enough to accommodate birds and bats. Meanwhile, if a tree remains 'undamaged' either by human pollarding or animal browsing, it is actually of far less wildlife value. 'Only 1% of oak trees younger than 100 year olds have hollows [while] … 50% of those aged 200–300 year old have hollows.'[16] Pollarding effectively speeds up the ageing process of trees like oaks, without causing any real harm. On top of that, altering the shape of our trees from having tall main stems to having multiple thinner shoots on a short, squat stem, makes them far more resistant to storm damage.

A pollard can produce a substantial new crop of strong poles every five years or so, nowadays much in demand by yurt makers. Pollards of oak or elm could also provide a supply of rot-resistant fence posts, which you would once have harvested for yourself right on your doorstep, from your own well-managed hedgerow.

But is it practical to hope for a wider revival of pollarding? It's more work for landowners and farmers, and it takes some skill, but if it were seen as a *crop* – which it is – grown for income, perhaps for woodchip biofuel, from trees that are more or less immortal on a human timespan, might we see more?

It seems hard to imagine that our forebears *ever* cut down a mature tree just for firewood, only for land clearance. Why kill the whole tree, when you can have firewood forever from a coppice or pollard in the form of bundles of thick poles – which is why in old novels and poems they are always burning not logs but 'faggots', meaning just such a bundle, a word that dates from at least 1300, according to the *Oxford English Dictionary*.

Rebecca Oaks and Edward Mills give a rough idea of just how much timber can be produced from a well-managed coppice wood, and this would apply to pollarded trees in hedgerows as well, which also produce numerous useful branches that can be repeatedly harvested. Depending on variables like how fast-growing the tree species was, how densely planted the 'stools' or tree bases were (and therefore how much sunlight each one received) and the local soil fertility, you might expect to harvest anything from 96 to 180 cubic metres of rods per hectare each coppicing cycle, which might be as frequent as every ten years.

And for *centuries*.

Most of the trees in medieval deer parks would have been pollarded, since coppicing would have left them open to constant browsing by the deer, but pollarding would produce a crop of timber and keep each tree growing far beyond its natural lifespan: in the long-term *reducing* rather than creating work, as you'd hardly ever have to replant your park.

During long hot summers when the grass was dry and brown and the animals were bleating or braying with hunger, the hedgerow pollards would still be green (thanks to their deep roots), and no doubt the shepherds and cowherds would then cut great swathes of branches from such trees – another rough and ready form of pollarding. I tried this myself in the summer of 2023, when the grass looked a little thin with the lack of rain in May and June, cutting big branches of grey willow from down by the river and throwing them in for our cows. They gobbled them up.

And I will never forget when I was travelling in Nepal, walking down a remote mountain road and seeing, to my considerable alarm, a huge shrub walking towards me, like Birnam Wood in *Macbeth*. It gave me quite a chill – until, as it drew closer, I saw a pair of small, elderly, but strong-looking legs underneath it. And then the Nepalese lady, the bearer of this colossal load of brushwood, eased herself upright slightly as she passed me by and beamed. She was just taking a fresh load of branches and cuttings to feed her cows.

A memorable experience.

The Joy of Rotting Wood

Quite a lot of a healthy, flourishing, biodiverse hedge will be dead.

Just as in an ancient woodland, dead wood is a crucial habitat. About 650 species of beetle in the UK alone require dead wood at some point in their life cycle.[17]

Yet another feature of the traditional hedgerow that is too rarely seen, along with the well-managed pollard, is a dead tree. If left standing, its debarked, bony white timber craggy in a winter sky like a natural sculpture is the one of our most important habitats – and an intact dead tree lying and rotting on the damp ground is equally as valuable. Both are too often tidied away, perhaps because of some imaginary 'health and safety issue', in a ceaseless quest to make the countryside more lifeless and boring.

Bare and leafless standing deadwood is absolutely the favourite perch of many a bird of prey, giving the best all-round view, as well as a popular song post. Ideally, any dead and decaying tree, but especially an oak, should always be left to rot away gently over many decades, when it is actually of more value than in its prime. Such an oak is quite naturally what a pollarded oak will eventually become in time, thanks to those pools of water previously mentioned. If there are too many mature trees beginning to dominate and shade a hedgerow, and too few dead ones, one trick would be to select some for debarking around the base, thus deliberately creating standing deadwood for a plethora of creatures. But the hands-off land manager or deliberate rewilder might actually find, as I have done, that wildlife will rapidly create standing dead wood all on its own, if permitted. We had some lovely young birch trees, which I was furious to find the grey squirrels had debarked about half-way up. But, of course, this didn't kill the birches; they just started sprouting again from below, while the dead timber above quickly became a home for beetles and is now a favourite feeding post for woodpeckers.

Although deliberate debarking can seem almost cruel to the instinctive tree lover, the results in biodiversity can be spectacular. Because nature likes few things as much as rotting wood. Especially beetles, whose larvae both feed on the fungi-softened wood-rot and hide out from predators in there as well.

Dead wood, leaf mulch, flowers and nectar and fruit are all essential to the rose chafer, for instance, a handsome beetle of woodland edges. There they find nectar in spring, especially that of their favourite, dog roses; their larvae feed on decaying leaves and roots, often living in the soil for several years. After pupating they spend another winter in deep soil or decaying wood over winter, finally emerging in all their glory the following spring.

This sort of life cycle is typical of many beetles countrywide – the increasingly rare noble chafer is another, requiring umbellifers in summer, old fruit trees in autumn and winter, but could be catered for by a well-laid hedge, as long as it includes plenty of rotting wood as well.[18]

The link between veteran pollards and decaying, hollowed-out trees is very clear in the Reverend Gilbert White's immortal *The Natural History of Selborne*. (The estimable cleric is also said by the *Oxford English Dictionary* to *possibly* be the first person in history – in 1763, to be exact – to mark a kiss in a letter with an 'x'.)

He notes in Letter XXVIII, of 8 January 1776, that some ancient pollard trees were regarded with tremendous, even superstitious, veneration by the people of the village and also records a curious custom to do with shrews and cattle.

> At the south corner of the *Plestor*, or area, near the church, there stood, about twenty years ago, a very old grotesque hollow pollard-ash, which for ages had been looked on with no small veneration as a shrew-ash. Now a shrew-ash is an ash whose twigs or branches, when gently applied to the limbs of cattle, will immediately relieve the pains which a beast suffers from the running of a shrew-mouse [ie. a shrew] over the part affected; for it is supposed that a shrew-mouse is of so baneful and deleterious a nature that wherever it creeps over a beast, be it a horse, cow or sheep, the suffering animal is afflicted with cruel anguish, and threatened with the loss of the use of the limb. Against this accident, to which they were continually liable, our provident forefathers always kept a shrew-ash at hand, which, when once medicated, would maintain its virtue for ever.

A shrew-ash was created by boring a hole in the trunk, a captured shrew being put in and then the hole hurriedly plugged up again, 'with several quaint incantations long since forgotten'.

Sadly, as White notes, this wonderful tree, very old, grotesque and hollow, and surely a habitat for many owls, bats and others, had been chopped down by the previous vicar some twenty years before, 'regardless of the vain remonstrances of the by-standers, who interceded in vain for its preservation', although not perhaps for entirely ecological and twenty-first-century-type reasons.

Hedge Flowering Plants, Shrubs and Climbers

As well as trees and shrubs of all sorts, the hedge is also made up, of course, of flowering plants.

And here again, we're reminded that our ancestors really did know the value of everything in the hedgerow, including things that we now consider noxious weeds and spray to death with carcinogenic chemicals or even kill off with flame-throwers.

The humble stinging nettle, for instance, was once considered so valuable both for food and for fibre that a record cited in *Hedges* by Pollard, Hooper and Moore speaks of '8d for nettles in the garden sold this year', at Ely's manor at Lakenheath in 1429. In fact, nettle fibre has been used to make clothes since at least 4000 BCE, and today you can still buy Italian designer jeans made from nettles.

Another one of my favourite hedgerow 'weeds' is cleavers, or goosegrass, partly because it is so detested by tidy gardeners. Even the Royal Horticultural Society, disgracefully, recommends eliminating it with Roundup, 'probably carcinogenic' according to the WHO, and widely used the world over. Another gardening website recommends, under 'Natural Control', that 'flame guns are also effective'. Obviously, the word 'natural' is used with some latitude here.

'If it feeds the bee, it feeds thee,' said our wiser forebears.

This hated but lively little climber has over eighty different colloquial and regional names, including stick-a-back, sticky willy, bobby buttons, gollenweed, sweethearts (because of its clinging habit), kisses, claggy meggies and robin-run-the-hedge.

Like nettles, docks and so on, goosegrass is a sign of fertile soil. They 'live on phosphate, and mankind is a phosphate-accumulating animal'. The phosphate we accumulate – by farming, manuring, burning firewood and spreading wood ash, etc. – endures for centuries, as in the lost village of Little Gidding in Huntingdonshire: deserted and forgotten by 1640, yet still a place of 'ear-high nettles'.[19]

Goosegrass is rich in vitamin C, among a host of other healthy stuff, and, though it looks unpromising, sautés really well in springtime in some butter. Geese love it, obviously, hence the name, and so do our cows. Over the course of April and May, I watched them fatten up beautifully while gorging on the stuff, never making that baleful mooing that cows do when they're hungry, have stomachache or are otherwise less than perfectly contented cows. Their coats gleamed glossy in the morning sunlight, tinged with russet, and when Nick, their farmer-owner, came to look them over, he grinned with satisfaction. They were doing better than another lot of the same age he had on different, neater fields. I reckon it was (partly) the goosegrass, though I'm not sure Nick was entirely convinced.

The plant is actually in the *Rubiaceae* coffee family, and later in the year you can roast the seeds to make a low-caffeine substitute or squeeze the juice from the stems to curdle milk for cheese making. It's also loved by one of our most spectacular moths, the hummingbird hawk-moth (*Macroglossum stellatarum*) for which it's a larval foodplant. Roundup the goosegrass, and you will never get to see one of the UK's most stunning insects.

Other of our hedgerow bushes and climbers are more aesthetic though, I must admit. Bindweed can be stunning in full sunshine (it's rarely found in the heart of an ancient woodland), with its big tropical-looking flowers and one of the best of all old country names, 'Granny-jump-out-of-bed'; black bryony (our only native yam); and hops for beer, of course. Redcurrants, raspberries and gooseberries can also be found running wild in hedges, and the lovely traveller's joy (*Clematis vitalba*), although only on chalk. Hedgerows in chalk uplands will also be rich in other shrubs more rarely seen in the lowlands, including purging buckthorn, guelder rose, wild privet, spindle and wayfaring tree. Guelder rose's bright red berries are also technically edible – but you wouldn't really want to eat them. One authority says they taste like 'wet dog'.

Above all, there's honeysuckle. It looks gorgeous, smells gorgeous, the flowers are loved by butterflies (including the now rare white admiral), the autumn berries are eaten by thrushes, warblers and bullfinches, and dormice even like lining their nests with honeysuckle bark. In fact, they love the bark for bedding so much that they will travel twice as far to gather it as any other material.

Honeysuckle also attracts some of our most stunning nocturnal moths thanks to its trick of emitting a still stronger fragrance after nightfall as the moths take flight, including the pine hawk-moth and the convolvulus hawk-moth. If you have honeysuckle in your garden or a hedge near you, spend a few minutes at dusk in summertime looking for any of these colourful heteroceran arrivals.

And certainly, if you're planting a new hedge, don't forget the climbers: throwing in a few handfuls of blackberries, bryony and honeysuckle berries in autumn might work as well as anything.

Nutrient-rich hedgerow margins in the lowlands, such as the Underhill hedge, will feature abundant cow parsley and hogweed, the latter loved by cows just as much as by pigs, and generally crowded with beetles as well, especially handsome vermilion soldier beetles, usually to be seen in courting couples

– so much so, indeed, that they are jocularly referred to by entomologists as the 'hogweed bonking beetle'.[20] Hogweed is also known as humpy scrumples, cow bumble and, my favourite, pig's flop.

More delicate residents of the hedge might include bee-friendly hedge woundwort, selfheal, bugle and ground ivy – present at the Underhill hedge – while damper ditches may bring in willowherb, food for the spectacular elephant hawk-moth.

Ground ivy is an important but often overlooked little flower which is adored by bees out early foraging. Old country names like hedge-maids and creeping Charlie betray its long, probably millennia-old, association with hedgerows. It's a woodland-edge plant and was used in the past to clarify ale before hops were widely used, as well as for preserving the ale for longer, hence other old names alehoof and tunhoof – a tun being a barrel containing 216 gallons of ale. Shakespeare has Prince Henry fat-shaming Falstaff, in *Henry IV, Part I*, by calling him 'a tun of a man', though it's not as rude as his later insults: 'that bolting-hutch of beastliness, that swollen parcel of dropsies, that huge bombard of sack, that stuffed cloak-bag of guts'.

Shadier hedgerow margins will often feature wild garlic, one of the very finest of wild foods, and the handsome foxglove. There's a rich strain of folk knowledge connected with this flower and the heart-disease medicine digitalin. For it was in 1775 that Dr William Withering learned from an old gypsy woman a remedy for dropsy, or oedema, which included foxglove as a key ingredient. Foxglove taken wrongly is incredibly poisonous – just two or three leaves will stop your heart – but it also now forms the basis of digoxin, a key treatment for conditions like atrial fibrillation.

Another hedgerow plant said to have useful properties was the otherwise deadly poisonous hemlock (*Conium maculatum*), which of course did stop poor Socrates' heart. But if you make a kind of poultice or pulp of the leaves, according to the great seventeenth-century English herbalist, Nicholas Culpeper, it can relieve a distressing problem, for when 'applied to the privities, it stops lustful thoughts'.

As far as I know, this has not yet been subjected to serious clinical trials, but clearly medicinal plants of the hedgerow still have much to offer.

Since we have descended to earthier matters, lords-and-ladies (*Arum maculatum*) may have been unforgettably described by Thomas Hardy as resembling 'an apoplectic saint in a niche of malachite', but traditional country names are rather more blunt. Willy lily and dog's cock are just two, while the alternative 'polite' name, cuckoo pint, is really a shortened and sanitised version: for pint read pintle, i.e. penis. The name lords-and-ladies is too a Victorian way of veiling what Richard Mabey calls the 'great catalogue of vulgarities' that surround this unignorably phallic plant. The berries are poisonous, but Dioscorides says they 'stir up affections to conjunction when drunk with wine'.

Hemlock is one of the umbellifer or carrot family, along with cow parsley, hogweed and so on. Some are edible, some are deadly poisonous, and some really aren't worth the trouble. If you dig up a wild carrot hoping to find a juicy and delicious orange root, you will be sorely disappointed. You'll just find a thin, white wispy tendril of a root, too small to bother with; but you should then pause and say thank you to those nameless generations of early farmers who developed the modern carrot from this unpromising start.

All of the umbellifers produce nectar loved by insects, and later abundant seeds for birds, so it's best to leave them where you find them. The various species together can provide a crucial succession of nectar from March through to November, their domed heads crawling and humming with hoverflies, longhorn beetles, soldier beetles and more. It isn't just nectar either: insects also eat high-protein pollen. Hoverflies require a good diet of pollen 'if the ovaries are to mature and eggs be laid … Nectar can be regarded as a "fuel" source for immediate energy requirements while pollen, rich in protein, is a body building or germ cell ripening food.'[21]

You can usually spot the first glossy leaves of cow parsley appearing soon after Christmas in sheltered hedgerows in the south, along with the snowdrops, while in *Hedges*, the authors actually observed cow parsley *in flower* in January![22]

Closely following in early spring come two lovely little yellow flowers, the primrose and the celandine, specifically the lesser celandine, gleaming glossy gold in the grass, sometimes as early as mid-February on a sheltered south-facing bank. Gilbert White noted its appearance on 21 February. Celandine derives from the Greek for swallow, *chelidon*, 'bycause that it beginnith to spring and to flowre at the coming of the swallows and withereth at their return', says a sixteenth-century herbal. Lesser celandine is also known as pilewort, though whether it is effective for this particular affliction is unclear. It used to be called *Ranunculus ficaria* by botanists, the '*Ranunculus*' bit associating it with frogs, with which it often shares a damp, marshy habitat. But nowadays it bears the rather duller name of *Ficaria verna*.

Wordsworth loved this first little flower so much that he asked for one to be carved on his gravestone in Grasmere churchyard, but, unfortunately, he got a greater celandine instead. Surely after over 170 years, it's time for this to be corrected?

Another humble but important flower of the hedgerow is the diuretic dandelion, *pissenlit* in French, whose springtime leaves you should definitely eat, even if they do make you wee a lot. Later, it's marvellous to watch the goldfinch like an acrobat on a trapeze, hanging from the waving stalks to feed on the dandelion clocks as early as May.

Such is the wonderfully varied and complex plant life of our hedgerows, from elegant, malleable and forgiving hazels to tough and thorny structural staples like hawthorn and blackthorn, to veteran oaks that have seen centuries pass, to the multitude of flowers, from the common (but invaluable) bruisers like brambles and nettles, goosegrass and hogweed, to humble snowdrops and celandine, as well as those hedge flora that indicate *probable* antiquity, like bluebells, yellow archangel and wood anemone: forest flowers of the woodland edge.

Most poignant are the rare glimpses of occasional cornfield flowers that you might see on dry, bare hedgerow banks, long since driven from arable land by agricultural (i.e. biocidal) spraying: cornflowers and corncockles, for instance. Other arable flowers, however, like darnel, thorow-wax, also called hare's ear, and lamb's succory, a kind of chicory, are regarded now as extinct in the wild.

There are more rarities you might find in their last refuge in hedges or even busy roadside verges, even *those* being more hospitable than modern farmland. Stationary in a traffic jam on some unlovely dual carriageway, you might still see some wood cranesbill in the verge beside you or globeflower, or the rather aptly named melancholy thistle.

Other Elements: Margins, Soil ... and Lichens

The precious grassy margins of hedgerows can also be very rich in wildlife, all part of the same edge habitat. Yet they are still too often cut too close to the hedgerow or 'sprayed off', even though the science is now in on how they can directly benefit farmers. A 2016 study from Switzerland on the effect of hedgerows' accompanying wildflower strips found that wheat yields were higher up to ten metres into the field where such strips were present, while damaging aphid numbers on potatoes were down a massive 75 per cent.

The total area of our roadside verges alone adds up to over half a million acres.[23] Twenty of our fifty-five mammal species, all six reptiles, five out of six amphibians, and at least sixty butterfly species have been recorded on these verges, even with traffic going past.

Even more unconsidered is the soil *within* the hedgerow: a linear compost heap, as complex and teeming with life as the soil of ancient woodland, including things like ground-nesting spiders, false scorpions, millipedes and springtails. There can also be about one hundred million bacterial and fungal colonies per square centimetre of decaying leaf. And feeding on these, there could by any of the 250 species or so (we're still not exactly sure) of springtail native to Britain within a hedgerow's humus.

There will be numerous other organisms living lives of incredible complexity and interdependence, which science is only just beginning to unravel. As George Monbiot has pointed out, it was only in 2016 that scientists first came up with a satisfactory definition of what soil actually is! 'The soil is a living, four-dimensional natural entity containing solids, water (or ice) and air ... and consists mostly of a structured mixture of sand, silt and clay (inorganics), rocks and organic material (dead and alive).'[24]

This is a vast subject, then, rather beyond the scope of this book, but suffice it to say that hedges nourish and protect soil to a great degree, and the soil in turn stores vast amounts of carbon dioxide (see chapter six), as well as being home to innumerable creatures.

And finally, a note on one vanished feature of our hedgerows, a feature lost for so long that we no longer even see it: what ecologists called 'shifting baseline syndrome'.

Ancient and medieval hedgerows, it seems, like ancient woodlands – not just our lost rainforests, but *all* our native woodlands – might once have been absolutely festooned in mosses and lichens.

For example, one lichen, tree lungwort (*Lobaria pulmonaria*), requires virtually zero pollution, and so today most of us encounter it in our countryside as often we do a wolf or a lynx or an apple bumblebee. Yet regarding this freakish lack of lichen in our damp, rainy islands, 'Anyone returning from the Middle Ages would immediately comment,' observes Oliver Rackham, for the great old oaks of those days would have been liberally shrouded and hung with tree lungwort and all the other lichens.[25]

But hedgerows and coppices? Weren't they cut too often for lichens to develop? Apparently not. For in the wattle and daub old houses of Suffolk, reports Rackham, one can find hazel rods with shreds of medieval moss and lichen still attached.

While we remain addicted to cheap food and the ongoing eco-disaster of intensive farming as the favoured way to feed our vast and ever-growing population, our hedgerows offer at least some respite from the onslaught on our wildlife. Deteriorating though many are, much in need of better management, they often remain safe refuges for invaluable old features of our landscape, like pollarded oaks and rare plant species like spreading bellflower, Deptford pink, limestone woundwort and lady orchid, all now officially classified as Endangered by the International Union for Conservation of Nature but still – just possibly – surviving in a sensitively managed hedgerow near you.

The names alone make you want to write a kind of poem, the same way the richly evocative village names of Dorset inspired John Betjeman when he wrote the opening line:

Rime Intrinseca, Fontmell Magna, Sturminster Newton and Melbury Bubb…

My hedgerow rhyme would go something like:

Betony, bryony, meadow clary, woodruff and wild service tree, Babington's leek, mountain melick and narrow-leaved everlasting pea…

I hope it's not an elegy.

CHAPTER THREE

Nature's Hum

Hedges and Insects

Now Summer is in flower, and Nature's hum
Is never silent round her bounteous bloom;
Insects, as small as dust, are never done
With glitt'ring dance, and reeling in the sun;
And green wood-fly, and blossom-haunting bee,
Are never weary of their melody.
Round field and hedge, flowers in full glory twine,
Large bind-weed bells, wild hop, and streak'd woodbine,
That lift athirst their slender throated flowers,
Agape for dew-falls, and for honey showers

John Clare, *The Shepherd's Calendar*

T he traditional British hedgerow in high summer, especially if south-facing – and so beautifully evoked here
by John Clare, the greatest of our nature poets – will be
such an invertebrate haven, so loud and brimming with insects
that it will seem the perfect embodiment of what W.B. Yeats
called 'life's own self-delight'.

An east-facing hedgeline will see an early stir of birds and
then insects, while a southerly will be a-flutter with butterflies
in high summer, especially if natural curves and scallops have

been created or tolerated, offering extra shelter and warmth. To the west, birds will sing their sunset song from the highest perch they can find, but the damper northerly sides are just as crucial, the favoured retreat of shade- and moisture-loving slugs and snails, and the amphibians that feed on them.

A hedgerow in Britain can support up to 1,500 invertebrate species across seventy families from hawthorn, associated with 209 species, to the waxy, prickly holly at only ten (although one of these is the lovely holly blue butterfly).[1] Another study found slightly fewer: fifty-one families in the hedges studied, of which 31 per cent were beneficial predators and 6 per cent parasitic.

But even these impressive numbers may have to be revised upwards, and the eco services of the hedgerow too, thanks to the heroic labours of Dr Rob Wolton and his ninety metres of Devon hedge, which he has been examining in detail for years.[2] He started surveying this one hedge in immense detail back in 2011, aiming to count every single species of plant, animal and fungus that he found and expecting initially to spend twelve months on the project. Today, he is still at it. And he isn't even counting microbial life, only what he can see with the naked eye.

He has so far 2,070 species of plant, animal, lichen and fungus, with many more awaiting identification, among them 1,700 species of insect and rising. Overall, Dr Wolton suspects his hedge may harbour some 3,000 species. If he were to include microscopic organisms, the final number might be mind-boggling.

Pests and Predators: A Fine Balance

Among a hedgerow's most important denizens on farmland are the voracious predators like ground beetles and rove beetles, which overwinter in hedgerows before moving out in spring to feed on aphids and slugs' eggs in the crops, both aphids and slugs being highly destructive to farmers' crops and gardeners' produce. Other helpful insect predators, like hoverflies, lacewings and ladybirds, can consume vast quantities of aphids and/or their honeydew, which contains nutritious sugars, proteins and amino acids.

Aphids can reproduce by parthenogenesis – literally, virgin birth – which is why one or two of them can turn into a thousand seemingly overnight; or more mathematically, why just one can turn into six hundred billion in a single season under optimal conditions. Fortunately, nature is one vast and generally efficient system of checks and balances, and so many other insects eat aphids that things tend to even out over the course of a year. Nature does indeed have a plan, and it's rather more sophisticated and complex a plan than spraying broad bean plants with soapy water in the hope of dispatching the critters.

It was the same on our cherry trees, as I learned recently. At one point in spring, they were absolutely covered in sap-sucking aphids and not looking entirely happy about it. But, as Harry James points out in his dissertation, aphids are actually a keystone or foundation species, feeding so many other species higher up the food chain, that an abundance of them is a healthy sign. A week or two later, their many insect predators had emerged from our numerous log piles, hedgerows and thickets and gobbled the lot.

In other words, although the insect world displays a fecundity and numerical abundance that is both marvellous and slightly terrifying, overall, it works very well – especially if we give it a general helping hand, rather than trying to spray one isolated part of it to death. And tending hedgerows is a crucial insect-friendly strategy, as we shall explore in this chapter.

Carabids – those aforementioned ground beetles – are very often found in hedgerows, even though they are technically a woodland species and highly predatory, with 'a well-developed hedge base and the presence of a ditch' greatly enhancing the population of predator species.[3] For comparison, broad-leaved woodland, so stoutly and rightly defended as a wildlife refuge, only supports 20 per cent of our beneficial predators. In other words, hedges are a food grower's greatest friend in terms of potential pest control.

Taking a wider view for a moment, in California's Central Valley, one of the most intensively farmed zones on the planet, new hedgerows have been planted, and invertebrate visitors to 'hedgerows tailored to the needs of the regional pollinator fauna' have been shown to include 173 species of bees and hoverflies, from 50 different genera. The results have been excellent because healthy bees create healthy croplands. Exactly the same principle holds in Britain.

Balanced, productive farmlands, their hedgerows broad and venerable, teeming with invertebrate life as in John Clare's day, and then with all the other mammals and birds that feed on them, are tragically rare today, just another facet of our skewed reality. But the reconstituted hedgerow at Underhill gives a glimpse of how farmland boundaries might look again, with sufficient encouragement and incentive. It is immediately noticeable, observing the Underhill hedge on a May morning, how vital and busy is the interaction between such a hedgeline and the adjacent grassland. Breeding birds, in particular, dart down into the waving grasses to gather beakfuls of leather-jackets, the cranefly larvae so abundant in such healthy meadowlands. A typical insect larva will be about 20 per cent protein and a further 12 per cent fat, so just the thing for a growing fledgeling.

Our 400,000 kilometres of hedgerows nationwide, ideally such an oasis of life in an otherwise denatured and often eerily silent countryside, are often anything but, thanks to poor management, brutal cutting regimes, neglect and the general deterioration of the ecosystem around them. If a field alongside a hedge is still being drenched in biocidal chemicals multiple times every year, the hedgerow will hardly flourish with life.

So, first of all: how bad are things now?

The End of Insects Means the End of Everything

In 2021, Dave Goulson published a book called *Silent Earth: Averting the Insect Apocalypse.*

The title was an obvious tribute to the Rachel Carson classic, *Silent Spring.* But surely the subtitle, the 'insect apocalypse', was a little hysterical?

Goulson is professor of biology at the University of Sussex, and has been studying insects, especially bees, for decades. And in his book, based on abundant research by himself and others around the world, he concludes that insect numbers have fallen by some 75 per cent from healthy, normal, pre-industrial levels. If the decline continues at this rate, and is linear (which it probably isn't; it may even accelerate) – insect numbers may have fallen so low as to trigger worldwide ecological collapse within twenty or thirty years.

Some scientists raised objections to Goulson's assertions.[4] Meanwhile, others have agreed with Goulson that our flying insect population is down some 60 per cent in just twenty years. Two conservation charities, Buglife and the Kent Wildlife Trust, persuaded members of the public to do a bit of citizen science and count up the number of bugs squashed against their car number plates, and they were able to compare this to similar data from 2004. Sure enough, it grimly confirmed Goulson's worst fears. Numbers were down by 65 per cent in England and 28 per cent in Scotland.

We also know that heathland and chalkland butterflies have dropped some 68 per cent since 1976, with half of all butterfly species now at risk of extinction, according to the UK Red List, published and regularly updated by the Joint Nature Conservation Committee. This is a national disgrace.

In this, as in so many green issues, if one reads first one set of experts and then another set of experts opposing the motion, one can be left in a kind of fog-bound impasse, concluding with a layman's helplessness: 'I just don't know; it's all so complicated.'

This is the great danger of trusting the science and *only* the science. It makes us passively, absurdly dependent upon scientists to tell us the truth. And yet they frequently disagree with each other. Meanwhile, without their knowledge and their apparatus of peer-reviewed research papers, verified data and rigorous scientific method, we can apparently be sure of nothing at all.

This is ridiculous, of course. Scientific method only arose, at the very earliest, around 1600 CE, which would imply that before that, for some 200,000 years or so, all those countless generations of *Homo sapiens* who successfully grew to adulthood, had children, hunted, gathered and grew crops, fed their children, treated their own ailments, cured their illnesses and fought off wild animals, effectively knew nothing.

While science is invaluable, a wonder of the modern world, it isn't the only way of knowing. What about our eyes and ears? What about talking and listening to people with a love of nature and long memories?

It's these simpler forms of knowing that can confirm that Dave Goulson, and many others who agree with him, are surely on the right lines, and the sceptics are wrong. In support of the insect apocalypse theory, we can adduce: the massive and very well-known decline in squashed insects on our car windscreens; the number of people who say the countryside in summer 'no longer hums with insects like it used to'; the dramatic fall in many species of insectivorous birds; and so on. Pure science may dismiss these things as 'merely anecdotal', as does the critique of Professor Goulson. But 'empirical' would be a better word. If the theory of an ongoing insect apocalypse chimes precisely with my own experience of walking through the English countryside in summer, crossing field after field of arable crops, and seeing barely a single insect, then I can be fairly sure Goulson is right. Even more so if I talk to an elderly local farmer, now in his eighties, who laments the astonishing decline in butterflies and birds since his own boyhood.

So toxic, indeed, so hostile to life itself is much of our farming land, apart from that single mono-crop being cultivated, that the University of Plymouth recently found that hedgerows and verges facing the roads – with all their noise, dust, wind, diesel fumes and so on – are far richer in flowers and bees than the other side, the supposedly quiet side, the sheltered side, facing into the farmer's field.[5]

And we're only beginning to learn about the insects that remain. Only very recently, we discovered that bumblebees are electrostatically aware. A bumblebee visiting a negatively charged flower cancels it out, so the next bee literally fails to feel much of a buzz and knows without further investigation that there would be no point expecting any nectar. Time to go on to the next flower. Bees can even smell a prior visitor's cheesy feet.

The insect world is a world of astonishing beauty and complexity, and we are erasing it just as we are beginning to understand it.

However, there are numerous things we can do to fight back against this decline, without waiting for some kind of permission from the scientific consensus – which could be a whole new kind of 'learned helplessness'.

One is to try and inform people. This can be hard work. To my astonishment, I have talked to several friends, educated, successful, readers of *The Times* and the *Financial Times*, who have given a shrug and said they can't see what difference it would make. 'Why do we need beetles anyway?' queried one (an alumnus of Oxford University, no less).

This can be almost as disheartening as the insect collapse itself. After all these years, have they still not absorbed the most basic principle of life on earth? That all species are interdependent, and couldn't survive without each other? Apparently not.

So, we might alert some people, but not all. Meanwhile we can get on and do some practical things to help. No one can stop us rewilding our gardens or planting trees where suitable, and perhaps even more importantly, creating (or recreating) wetland by the simple act of 're-bogging'. I managed to re-bog about half an acre of our old water-meadow by the elementary act of diverting a rainwater drainage ditch running down from the field above, with our neighbour's agreement. The result was a stretch of year-round squelchy marshland which in summer saw an absolute explosion of invertebrate life, and imminently, I hope, amphibians.

And then there are, as William Wordsworth puts it, 'those little lines of sportive wood run wild', those insect habitats par excellence…

Hedges!

Hedges for Insect Habitat

In negative, a landscape of vast fields without hedges or any other windbreak or nectar source is a disaster for insect life, and therefore all other wildlife that is dependent on it.

Hedges acting as windbreaks also lessen soil erosion, of course. A hedge of 40 per cent permeability will reduce wind speed by at least 20 per cent up to seventy-two feet into the field, in the case of a six-foot-high hedge, which may sound like a marginal difference – but when you're the size of butterfly, let alone a gnat, marginal differences can be huge.[6]

A study in Canada found that field size was also crucial to insect abundance: for butterflies, hoverflies, bees and ground beetles, the smaller the better.[7] A key conclusion of the study was simply that: 'Reducing field sizes could be an effective conservation action in agricultural areas.'

And to reduce field sizes: just plant and maintain hedges.

Some farmers and landowners mutter about the cost of doing so. But what of the costs of not doing so?

One study suggests that the ecosystem services of insects and other invertebrates are worth $57 billion per annum to the US. Another study reckons the value of wild bees in the UK at £3,000 per hectare, or £651 million a year.[8] There are some 3.2 trillion bees at work worldwide and they cannot be replaced.

Yet for some, extinction is already here.

The blue stag beetle (*Platycerus caraboides*) is now extinct in the UK. In fact, while we only have three remaining species, a record by a nineteenth-century parson (of course – who else?), Reverend F.W. Hope, an Oxford undergraduate from 1819 to 1822, tells a different story. Collecting beetles in Wytham Woods, Oxfordshire, he noted five different species of stag beetle. 'These are of extreme interest, because they include two species now regarded as long extinct in Britain: the largest species of burying-beetle in Europe, *Necrophorus germanicus*, and a stag-beetle, *Platycerus caraboides.*'[9] There are further records of now extinct stag beetles from Bristol and Windsor, and no doubt many more went unobserved.

It is just this sort of unnoticed ebbing away of our wildlife that causes most dismay. Our remaining three stag beetle species are becoming rarer by the year: the European stag beetle (*Lucanus cervus*) with its handsome mahogany antlers and wing casing; the smaller, all-black lesser stag beetle (*Dorcus parallelipipedus*), found throughout England as far north as Nottinghamshire, but which particularly likes living in Watford; and the rhinoceros stag beetle (*Sinodendron cylindricum*), with its distinctive rhino-type horn. Should we not at least preserve what coleopteran wonders we have left?

Another insect gone forever, whose absence we no longer even notice, is the spotted sulphur moth (*Emmelia trabealis*), once resident on the Norfolk–Suffolk border. 'Its larval food-plant was (and presumably still is in the rest of its European range) Field Bindweed, *Convolvulus arvensis*, and it flew in a single generation from May – July.'[10]

There's still plenty of field bindweed around, of course, but the spotted sulphur is gone from an East Anglia now intensively farmed. And it was a staggeringly beautiful creature.

Dung beetles are another whole cluster of species in serious decline, killed off by farmers giving worming medicine to their cattle, which then passes through into their dung and kills other insects as well. Yet the multiple labours of dung beetles, recycling manure, enriching and aerating the soil and so on, are estimated to be worth £367 million per annum to UK cattle farming through the provision of eco-system services.[11]

Dung beetles have been breaking down and recycling the manure of large ruminants and grazers for around 65 million years; generally little appreciated until they disappear, when the results can be disastrous. Witness the first introduction of cattle to Australia, and a dung beetle population only evolved to deal with the dung of marsupials: result, thousands of miles of non-rotting manure.

Other beetles help to break down and recycle dead wood, and indeed I have long wondered whether the massive decline in beetle numbers across the country over the past two centuries or more, and particularly in the past fifty years, has meant that fallen branches and timber now decay *much more slowly* than they used to. As a result, farmers, landowners and gardeners are always clearing and burning wood; otherwise, as you frequently hear it said, 'it just sits there'. But beetles are experts at disposing of your dead wood for you, if you give them a chance, often in alliance with fungi. In fact, beetles can't really begin to digest wood at all until fungi have done some of the softening up first, and so incredibly, some species of beetle, such as ambrosia beetles, take their favoured species of fungi with them when moving to a new site. Is this yet another example of animals farming – wittingly or unwittingly – like jays planting oak trees, squirrels planting hazels and ants farming aphids?[12]

An Ecosystem to Revive

I would suggest there is also a kind of 'reverse' way of appreciating what we cannot actually see, namely, how many insects there *aren't* in our countryside anymore. And that's from considering past numbers of insectivorous birds.

A single breeding pair of swallows and their offspring require a staggering one million insects per season. So, where there were once such huge numbers of swallows forty or fifty years ago – let alone one hundred – just imagine what an accompanying abundance of insects there must have been.

But how much have swallows declined – along with house martins and swifts? Many wildlife bodies are cautious, and suggest 40 per cent, adding that their numbers do fluctuate anyway. But talk to any observant country dweller with a long memory, and they will tell you that swallow numbers have fallen catastrophically since they were children and remember the skies alive with them – as I do myself, just from the 1970s.

Older nature writers talk of flocks of swallows and their relatives in their *tens of thousands*:

> In the case of the Swallow (*Hirundo rustica*), and the House-martin (*Hirundo urbica*), the flocks thus formed consist of thousands. We have seen clouds of swallows … thousands were dashing about in every direction through the air … Even more multitudinous are the flocks of house-martins, which may be now observed along the banks of the Thames from Hammersmith to Richmond … the roofs of the adjacent houses swarm with them, clustering together like bees. They form, as we can testify, a cloud in the air.[13]

What a sight it must have been! And this at a time when London famously suffered from dense coal smoke and smog. Yet there must have been vast armies of insects even in London to support such populations. Mere written descriptions like these aren't admitted as evidence by modern ecologists, who like exact figures, which tend to date only from the 1960s and 1970s – but to me they are telling indeed and suggest a decline far greater than realised.

A single of flock of thousands of swallows would require *thousands of millions* of insects to raise their young in a season; and there were many such flocks.

Such examples of the very long-term decline in our wildlife, especially in insect-eating birds, can be multiplied almost ad infinitum.

'In the summer, it is difficult to be in any part of the country without hearing the cuckoo.'[14] An abundance almost unimaginable now.

Those energetic insect-eaters, spotted flycatchers, fell in numbers by 50 per cent between just 1965 and 1976 – the golden age of hedge destruction in the UK, one might mordantly say – and today are a rare sight in many parts, though described as quite common by Edwardian nature writers. And perhaps even more so William Cobbett, who calls them the 'bee-bird', in his *Cottage Economy*, and implies that they were so abundant in the 1820s that they constituted a serious nuisance to the beekeeper.

And so on.

It can be deeply depressing to ponder these descriptions of our lost countryside, and to realise how little nature there is left to us compared to even our grandparents' day, but the truth should be faced.

Another way in which farming is skewed against wildlife and wider ecosystem health is in its love of the monoculture, 'efficient' in the short term, but poor in the long term. An ecologically diverse environment is more stable *in itself* than a simplified one. And that unhedged, treeless, pond-free, tidy, seemingly well-managed and productive modern arable prairie is by definition teetering on a knife-edge, only kept going by a whole arsenal of chemical feeds, sprays and toxins.

Other types of perfectly productive human-made and managed farmland offer a striking contrast, especially something like an organic orchard. The orchard I have planted in a corner of our field, an unplanned muddle of cherries and damsons, apples and pears and medlars, has over the ten years or so of its existence given accommodation to all sorts of so-called pests and diseases. But I'm glad to say I held my nerve, rejected any idea of a chemical spray gun, maintained a Zen-like calm and did absolutely nothing, as I long ago learned to do from the wonderful *One Straw Revolution* by perhaps the world's only well-known macrobiotic Japanese rice farmer, Masanobu Fukuoka: the bible for the non-interventionist gardener.

And to date, only a single one of some thirty or so fruit trees has had to be taken out and removed as too sickly to keep. Other problems, like larvae of various sorts boring into apples, especially the prolific codling moth larvae, have actually become *less* of a problem over the years, and I strongly suspect this is because a healthy population of beneficial predators has now taken up residence in the orchard, for instance certain wasps and hoverflies, along with insectivorous birds, encouraged by the appealing habitat of a thick old hedge to the north, a much lower one to the south, which I planted myself, and plenty of rotting log piles. There is also an extremely good population of ladybirds, to be found on every single fruit tree for many months of the year, doing their busy bit to keep down fruit pests. The adjacent hedgerows have, I'm sure, been key to this, as has not spraying, and indeed barely intervening at all. Not having to expend energy fighting off insect pests means that the fruit trees are more able to resist bacterial and fungal attacks, and continue to flourish.

Experience in other parts of the world, particularly in Nova Scotia, confirms my own hands-on experience: orchard ecosystems are naturally much more stable than those of arable crops, largely because the habitat is more permanent and more diverse.[15]

If only farmers could be persuaded that farmland with more wild patches, including flourishing and structurally varied hedgerows cut on a three-year rotation, is far more sustainable and, ultimately, high yielding.

The damage that modern flailers do to the hedgerow otherwise, as a complete ecosystem in miniature, is horrendous.

And standing dead trees just get in the way so will usually be swiftly removed: another damaging 'simplification' of the landscape.

If a standing dead tree in full sun dries out completely, then decay may come to a halt for some years, neither beetles nor fungi able to devour it further. But nothing is wasted: the old holes and furrows of beetles, made when it was still moist enough, will now be colonised by solitary wasps and bees, the wasps even appointing one small tree burrow as a larder for flies and leafhoppers, another for their eggs, and so on.

In other words, maybe we shouldn't bother about those commercially made bug boxes, often somewhat overpriced. Just leave enough standing deadwood in our hedges and we will have the effective volume equivalent of millions of bug boxes.

There are other insect wonders that might inhabit a flourishing English hedgerow. If it includes a mature oak or a beech, then the hedge may be home to the largest aphid in the world, *Stomaphis quercus*, about the size and colour of a roast coffee bean, but still nowhere near large enough to seriously trouble the tree. These aphids are carefully tended by jet black ants who, like alpine shepherds, carefully carry the aphids up the tree in springtime to feed on the sap of fresh young leaves, and then bring them back down again in autumn to shelter in the thicker fissures of the bark.

If you have a slightly damper hedge, or a north-facing one, it may be home to a European nursery web spider, common throughout the UK and, we are proud to say, recorded as present at the Underhill hedge (see the table at the end of this chapter.) The nursery web spider has one of the weirdest courtship rituals in the animal kingdom – although many spiders are somewhat unusual in this regard, it must be said. During courtship, males present females with a 'nuptial gift' in the form of an insect wrapped in silk for her to eat. Until fairly recently, this gift was thought to protect the male from becoming the female's next meal. Research has shown, however, that the gift entices the female to mate and, what's more, the larger the gift, the longer she will mate for. (This in no way echoes the way that human males buy human females extravagant gifts in the hope of certain rewards, of course.)

All the same, just to be safe, quite a lot of male spiders have been observed 'playing dead' as well, while the female munches through her gift, in case she starts munching on him too.

You don't need to go to the Serengeti to see amazing animals. You just need a good thick hedge.

Hoverfly Heaven

Few insects are quite as beneficial in farmland, in orchards and in gardens as the amazing 120-wingbeats-per-second, 3.5-metres-per-second hoverfly. (The biting, blood-sucking horsefly, for comparison, can cover 4 metres per second, or about 9 mph: you're going to have run *quite fast* to get away from a determined one. And a hornet can do 13 mph: just so you know.)

Dead trees with pools of rainwater trapped between the boughs – or indeed pollards – may produce 'rot cavities' or 'hoverfly lagoons': high-rise breeding grounds for the very common hoverfly *Myathropa florea*, which produces a rather extraordinary-looking rat-tailed larva, the 'tail' actually being a long tube to allow it to breathe underwater in those rain-filled cavities. The hoverfly is also popularly known as the Batman hoverfly, because the pattern on its thorax roughly resembles Batman's mask.

Plants that can especially attract some of our roster of 276 native species of hoverfly include both yarrow and hogweed, both of which might be found in a well-tended grassy margin or hedgerow and therefore provide huge benefits. Indeed, they may be more important even than bees as global pollinators. Hoverflies pollinate at least 72 per cent of all food crops world-wide, they transport pollen over wide distances, including the four billion or so hoverflies that migrate into the UK every springtime from the Continent – and even more will then migrate back to the Continent in the autumn.

Yes, the UK is a net exporter of hoverflies, along with whisky, yachts and rubbish television.

Hoverflies, unlike bees, are also voracious consumers of pests. Their larvae consume some *ten trillion* aphids annually in the UK. A landowner or gardener who fussily 'sprays off' any hogweed or yarrow, thinking it's a weed, makes things harder for themselves, lessens or destroys the hoverfly population locally, and will reduce vital pollination services in any nearby orchards as well. One experiment found that strawberry yields increased 70 per cent when plants were visited by hoverflies compared to strawberries that weren't.[16]

In another experiment, strawberries pollinated by insects like hoverflies and those only self-pollinated or pollinated by the wind were photographed and compared. 'Only the insect-pollinated strawberry appears recognizable as a strawberry – the other two are so shrunken and misshapen they look like someone has gnawed away most of the plump fruit and tossed the remnants away.'[17]

Bees rarely forage more than two kilometres from their hive, but hoverflies cover hundreds of kilometres and migrate for thousands. They can even fly over one hundred kilometres of open water, 'transporting viable pollen over long-distances, thereby facilitating high levels of gene flow between plant populations that would otherwise remain unconnected. This, in turn, may have beneficial consequences for plant population health and fruit yield.'[18]

And what sort of countryside best supports hoverflies? A complex, natural-looking countryside rather than a simplified and modernised agricultural landscape. 'Woody elements such as *hedgerows* [my italics] are associated with increased local abundance of hoverflies and ecosystem services such as bio-control of crop pests, especially when connected to forest.'[19]

We have plenty of dense hedges and hoverfly-friendly umbellifers near our orchard – which grow of their own accord, of course, requiring nothing from me but to be left alone – and our fruit trees are never troubled by any explosions of pests. Some landowners also believe that dead wood can 'harbour pests' that might spread to crops or infect nearby healthy trees, but again this is false: the dead wood is just as likely to harbour beneficial predators.

You shouldn't even underestimate the importance of dead and hollow flower stems for insects, as green summer lushness turns everything to a crackly brown – and definitely don't burn them, thereby producing yet more unwanted phosphate, not to mention deadly particulates. Dead flower stems and crispy dried-out flower heads are another generally very overlooked habitat for all kinds of insects, which may hibernate in them over winter, including micro-moths, beetles and hymenoptera (the bees, wasps, ants and sawflies). Earwigs love to live in a nice dry hollow stem, while seemingly dead burdocks, teasels and knapweeds will be visited by goldfinches and other seed eaters long into winter. In teasel – a most statuesque plant in late summer – the pith is eaten by a tortrix moth and the seeds by a phalonidia moth.

Some insects will even partition a hollow stem, such as elder or bramble, into neat chambers, laying a single egg in each chamber, carefully provisioned in advance with a supply of pollen or nectar, and then seal up the stem again with mud, sawdust or resin. The larva will turn into a prepupa and then doze through the winter in a diapause – suspended development – until spring.

So, leave your cuttings and rakings in sun-warmed piles in quiet corners. They don't need burning and will bring many benefits, not least the many solitary bees that will re-emerge in springtime: a direct correlation has been proven between the abundance of solitary bees in the spring, the abundance of hawthorn berries the following autumn (the flowers having been pollinated by the bees) and the abundance of migratory thrushes like fieldfares and redwings in winter, feeding on those delicious ripened haws. None of this is remotely surprising to the keen observer of nature, of course, but it's good to have it confirmed by science. Fieldfares and redwings will also tidy up and devour any old, slightly rotten or even half-fermented apples for you – one of the reasons why they can get so damn noisy, I suspect, like a pre-Christmas drinks party in Wetherspoons.

Insects, Hedgerows and Modern Farming

Removing hedgerows is all part of the wider simplification of farmland that increases volume of food produced by the tonne yet lowers its nutritional value by the year: surely one of the reasons why modern human populations are now often defined as both obese and malnourished. Having eaten your 2,000–2,500 calories per day, if your body still 'knows' that it hasn't had enough magnesium or calcium, it will prompt you to keep eating – though unfortunately this may well be yet more quintessentially twenty-first-century-type processed-wheat-flour-and-palm-oil snax, invariably a kind of shiny beige colour, known colloquially to our police these days, I am told, as 'garage food'. They eat a lot of it – which might explain why they find it so hard to catch criminals.

Modern farming has a great talent for maximising short-term crop yields at the cost of seriously compromising soil health. Farmland soil today is constantly being harrowed, sprayed and ploughed – far more deeply and intensively than with the old horse-drawn ploughs of even a hundred years ago – as well as repeatedly compacted by huge machinery, and often left bare to leach out its goodness under the drumming winter rains. All of these quick-fix practices have a ruinously disruptive effect on the delicate relationships between plant roots and soil fungi, greatly reducing the absorption of nutrients: just one of the reasons why a cabbage today contains far lower levels of vitamins and minerals than one grown in your grandparents' day. Add in higher temperatures and more carbon dioxide in the atmosphere, and you have faster-growing crops that are still poorer in nutritional value.

But can wildlife-friendly hedgerows run counter to this and literally increase nutrient levels in our food? Yes they can – most directly by increasing pollinator numbers, among other things.

It has been found that '75% of globally important crop species are enhanced by animal pollinators, contributing up to 35% of total crop production.'[20] But there are more complicated things at work as well, because it's telling that: 'Hotspots for certain micronutrient deficiencies in human populations overlap with regions of high pollination dependence for these micronutrients.' This is a vitally important point to digest. Where insects are scarce, or have been eliminated, and therefore pollination levels are much lower, the nutritional value of the crop can actually decline. Humans dependent on those crops for food thus become more malnourished, even if eating the same bulk of food and calories, and so, 'Loss of pollinators could thus increase the burden of diseases due to micronutrient deficiencies in some populations.'[21]

Today, our agricultural land is just as pollinator dependent as it ever was, while there are fewer and fewer of these pollinators and other beneficial invertebrates, thanks to our ruinous agrochemical regimes, spraying any single crop up to sixteen times a year. We require more pollinators to feed a growing population, just at the time when we are ruthlessly exterminating them from our croplands.

Hence the absurd situation in Southwest China, where so many insects have been exterminated by indiscriminate pesticide use that farmers are now going up ladders and hand-pollinating their apples and pears: a grim example of how by dousing nature with regular poisons we only make more work for ourselves, not to mention poisoning ourselves, of course. We share 44 per cent of our DNA with honeybees, after all. If a certain insecticidal spray kills honeybees, it's unlikely to do us much good either. Dave Goulson also notes hand-pollination now beginning in Bengal and Brazil, as well as dramatic falls in insect-pollinated but intensively farmed crops such as French beans in Kenya and blueberries in Canada.

Yet by one extraordinary estimate, despite doing all this damage to the delicate web of insect life, only 0.1 per cent of pesticides reach their intended target. The rest is lost to run-off, but it doesn't just vanish: it goes on to damage other delicate and vitally important ecosystems nearby. It also ends up in our drinking water.

Of neonicotinoids, says Peter Melchett of the Soil Association, about 5 per cent goes into the crop, about 1 per cent is brushed off the outside of the seed as it's planted (and this dust can be so toxic it will kill bees on contact), and a mind-boggling 94 per cent ends up elsewhere: in the soil, in waterways, ditches and field margins, and onward to poison streams and hedge-rows. Hence the absurd outcome that, as the BBC's *Countryfile* has put it, 'The very areas of a farm that we and wildlife see as safe havens are now more poisonous than the chemically treated crops themselves.'

Even nectar-rich bramble flowers, so beloved of bees and butterflies, can now be toxic to them.

One after another, the pesticides come along and then get banned – DDT in the 1960s, aldicarb in the 1970s, now neo-nicotinoids. But it doesn't matter how many lies Monsanto/Bayer or Syngenta spin us, the simplest understanding of nature and our common genetic descent tells us, as we've observed already, that if a certain manufactured chemical is formulated to be toxic to certain plants or insects, it will be toxic to many other living things as well. Duh.

You still sometimes hear it said in the mass media that our current catastrophic insect decline is 'still not fully understood by scientists', at the same time as we dump some *seventeen million kilogrammes* of pesticides on our countryside annually.

The same can be said of doggedly routine annual flailing as of pesticides. By reducing flower and fruit growth, farmers are again harming insects, since a diet rich in flower nectar improves insects' fitness, longevity and even fecundity.[22] Add to that the cutting and spraying of hedgerow margins, potentially such a rich site of wildflowers, and the damage is total. Even a marginal plant as humble as tussock grass, often overlooked by even the most ardent nature lover, let alone the average busy farmer, can host a range of beneficial insects like rove beetles, soldier beetles and ladybirds. Not that anyone needs to go around planting tussock grass: it will find its way there quite happily if you just leave a generous grassy margin alongside the hedge of at least two metres wide and *don't spray it.*

Even flies are crucial pollinators, something few people appreciate, usually seeing them only as buzzing nuisances. They pollinate onions, carrots and fruit trees, and, further afield, chocolate.

There are other reasons for the decline in our insects, of course, some of which are too rarely appreciated. Night-time light pollution is a disaster, for instance, and can diminish fruit pollination by as much as 13 per cent: a lot of fruit is pollinated by moths, who are deeply disrupted by artificial lights at night, and even more so glow-worms, whose numbers in England have fallen a terrible 75 per cent since just 2001 and must be well on the way to extinction within another ten years, when they will join the blue stag beetle, the apple bumblebee and the rest of the departed.

Another ominous reason for invertebrate decline – and one in which abundant, foliage-dense hedgerows can definitely play an important role in alleviating – is climate change.

Climate change is an enemy of insects in so many ways, including the obvious one of rising temperatures, and a disaster for bumblebees, which are among the best pollinators out there for things like tomatoes and berries. However, dense shady growth of any sort, from woodlands to thickets to hedgerows, offers just the kind of cool air-conditioned shelter amid rising temperatures that can help bumblebees enormously.

But it should be appreciated that, like mass tree planting, hedgerows are just part of the solution when it comes to mitigating climate chaos.

Another surprising way in which climate chaos threatens insects is that it produces faster-growing but less nutritious plants. (This will be the same for our own plant foods too, of course, threatening yet *more* human malnutrition in the future.)

This is important to understand since we hear a lot of glib assertions about how increased percentages of carbon dioxide in the atmosphere as a result of human activity will actually accelerate plant growth, and therefore increase crop yields, and so we'll all be better off, not worse.

Wrong in so many ways – not least in that we face climate chaos, not mere change or gentle warming, with increased storms, rainfall and general unpredictability, all of which make food growing and seasonal planning far harder.

But faster growing crops aren't a good thing, because rising levels of carbon dioxide are like junk food for a growing plant, especially if said plant is growing on ever poorer and poorer soil, which our intensively farmed soils always are. Carbon dioxide is accelerating their growth at the expense of their health, disease resistance, nutrient density, everything. It's a little like feeding a child an unlimited diet of chips and chocolate. Yes, said child will fatten up all right – but it'll hardly be healthy.

Researchers at the University of Oklahoma confirmed these suspicions with some vital research on tallgrass. They found that 'while the overall weight of the grasses had doubled over the past thirty years, likely due to the accumulation of CO_2, the plants' nitrogen content had declined by 42%, phosphorus by more than half, and sodium almost completely.'[23]

For grasshoppers, other insects and herbivores in general, this is making their meals 'more like iceberg lettuce than kale', according to Ellen Welti, the lead researcher. And for any insectivorous animal feeding on them, they will have to eat more insects to get the same nutrients, just when the numbers of available insects are declining. Thus, the negative cascades start piling up and the pyramid of life crumbles…

But hedgerows remain one redoubtable, consoling and eminently doable line of resistance to this gloomy tide. Henri

Clément, of the National Union of French Beekeepers, has specifically called for the restoration of hedges and mature trees in order to protect honeybee populations. Just as with human beings, honeybees do best and resist threats like the varroa mite, or the larger threat of colony collapse disorder, if they have the widest possible diet. Hedgerows can provide that – but hedge-free agro-plains of nothing but oilseed rape leave them, unsurprisingly, malnourished.

In 2013, in a genuinely meaningful and educational stunt, one Whole Foods store in Rhode Island removed all the food-stuffs that would disappear from our shelves in a world where our pollinators disappeared, or even reached critically low levels. Instantly, the store was bare of apples, avocados, carrots, citrus fruits, blueberries, green onions, broccoli and kale.

Earth's ecosystems have grown so dependent on the presence of insects because they have been around for some 480 million years, since the Ordovician Period. The web of actions, reactions and interdependencies between insects, plants and the rest of the living world is therefore enormously rich and complex. Even honeybees have been around for some 200 million years.

Incidentally, while on the subject of how long insects have been around, it is sobering in the extreme to realise that insects survived the previous five mass extinctions just fine. But this one is killing them. This tells us, in the bluntest possible way, just how bad this one could yet prove. Much faster, much more extreme and much worse, than any of the previous planet-wide collapses.

Oliver Milman gives a charming explanation of how much sugary nectar a bumblebee actually requires to fuel its multi-farious journeys and pollinations. 'If a human male devoured a Mars bar, he'd burn off the energy in around an hour; a bumblebee of an equivalent size would use the same energy in just 30 seconds.'[24]

Hedgerows are natural wildlife corridors, capillaries of humming insect activity running even through damaged landscapes, living examples of what the charity Buglife have called 'B-Lines' – but we need a lot more of them, and we need them to be better managed. Where this happens, there can be great success. At Underhill, there are now two wild beehives flourishing over twenty-five acres, and their honey goes untouched. They are an absolute delight in the summer – not least that happy, busy murmuring round the hives. If any bees will resist colony collapse, you feel these will.

The return of large-scale regenerative farming – with managed hedgerows at its heart – cannot come fast enough.

For a rich and inspirational vision of how our renewed and restored countryside might look, perhaps we should end this chapter where we started, with another wonderful description by John Clare:

Come, Queen of Months! in company
With all thy merry minstrelsy:—
The restless cuckoo, absent long,
And twittering swallows' chimney song;
With hedge-row crickets' notes, that run
From every bank that fronts the sun;
And swarthy bees, about the grass,
That stops with every bloom they pass,
And every minute, every hour,
Keep teazing weeds that wear a flower.

Appendix to This Chapter: Our Insect Survey of the Underhill Hedge

To give some idea of the abundance of insects that could be found in a mature, well-managed hedgerow such as the one at Underhill, we asked Marc Arbuckle, who has worked as the County Recorder for Wiltshire, to kindly come and do a survey for us.

It should be emphasised that this was primarily with a sweep net and only for one day in early summer. Marc pointed out that you would find many more species if you did overnight trapping, for instance, or used moth traps, and ground traps within the hedge for beetles and so on, or went right down into the rich humus layer inside the hedgerow to include mites and springtails.

Nevertheless, we were thrilled with the results, which Marc took right down to species level, showing what an old-established, foliage-dense and conservation-laid hedge can support:

Underhill Wood Nature Reserve Hedgerow Survey

Family	Scientific Name	Common Name
Acrididae	*Chorthippus parallelus*	Meadow grasshopper
Aphididae	*Cavariella pastinacae*	Willow-umbellifer aphid
Apidae	*Apis mellifera*	Western honey bee
Apidae	*Bombus pascuorum*	Common carder bee
Apidae	*Bombus terrestris*	Buff-tailed bumblebee
Apidae	*Bombus lapidarius*	Red-tailed bumblebee
Apidae	*Bombus pratorum*	Early bumblebee
Apionidae	*Perapion violaceum*	A seed weevil
Apionidae	*Protapion apricans*	Clover seed weevil
Araneidae	*Araniella cucurbitina*	Cucumber green orb spider
Argidae	*Arge pagana*	Large rose sawfly
Asilidae	*Machimus atricapillus*	Kite-tailed robberfly
Calliphoridae	*Calliphora vicina*	Common bluebottle
Calliphoridae	*Lucilia sericata*	Common greenbottle
Cantharidae	*Cantharis pallida*	A soldier beetle
Cantharidae	*Cantharis nigricans*	A soldier beetle
Cercopidae	*Cercopis vulnerata*	Red-and-black froghopper
Chernetidae	*Chernes cimicoides*	Common tree-chernes
Chrysididae	*Chrysis ignita*	Ruby-tailed wasp
Chrysomelidae	*Altica lythri*	Flea beetle
Coccinellidae	*Propylea quatuordecimpunctata*	14-spot ladybird
Coccinellidae	*Coccinella septempunctata*	7-spot ladybird
Coenagrionidae	*Enallagma cyathigerum*	Common blue damselfly
Coreidae	*Coreus marginatus*	Dock bug
Crabronidae	*Pemphredon lugubris*	Mournful wasp
Crabronidae	*Trypoxylon clavicerum*	A solitary wasp

Family	Scientific Name	Common Name
Curculionidae	*Phyllobius pyri*	Common leaf weevil
Curculionidae	*Phyllobius roboretanus*	A broad-nosed weevil
Curculionidae	*Sitona lineatus*	Pea-leaf weevil
Curculionidae	*Mononychus punctum-album*	Iris seed weevil*
Delphacidae	*Criomorphus albomarginatus*	A planthopper
Empididae	*Empis tessellata*	Common dance fly
Erebidae	*Tyria jacobaeae*	Cinnabar moth
Eriocraniidae	*Dyseriocrania subpurpurella*	Common oak purple (micro-moth)
Forficulidae	*Forficula auricularia*	Common earwig
Hesperiidae	*Thymelicus sylvestris*	Small skipper (butterfly)
Ichneumonidae	*Pimpla rufipes*	Black slip wasp
Ixodidae	*Ixodes scapularis*	Black-legged tick (deer tick)
Libellulidae	*Libellula quadrimaculata*	Four-spotted chaser (dragonfly)
Lycosidae	*Alopecosa pulverulenta*	Common fox-spider
Malachiinae	*Malachius bipustulatus*	Common malachite beetle
Miridae	*Capsus ater*	A mirid bug
Miridae	*Megaloceroea recticornis*	A mirid bug
Miridae	*Leptopterna dolabrata*	Meadow plant bug
Miridae	*Liocoris tripustulatus*	A mirid bug
Nemouridae	*Nemoura cinerea*	A stonefly
Nymphalidae	*Pararge aegeria*	Speckled wood (butterfly)
Oedemeridae	*Oedemera nobilis*	Thick-legged flower beetle
Oedemeridae	*Ishnomera cyanea*	A pollen-feeding beetle
Panorpidae	*Panorpa communis*	A scorpion fly
Phrurolithidae	*Phrurolithus minimus*	A spider
Pisauridae	*Pisaura mirabilis*	Nursery web spider

Family	Scientific Name	Common Name
Raphidiidae	*Phaeostigma notata*	Spotted snake fly
Rhagionidae	*Rhagio mystaceus*	Common snipe fly
Sarcophagidae	*Sarcophaga carnaria*	Greater worm flesh fly
Scarabaeidae	*Hoplia philanthus*	Welsh chafer (beetle)
Scathophagidae	*Scathophaga stercoraria*	Yellow dung fly
Scraptiidae	*Anaspis frontalis*	False flower beetle
Scutelleridae	*Eurygaster testudinaria*	Tortoise shieldbug
Sialidae	*Sialis lutaria*	Common alderfly
Stratiomyidae	*Chloromyia formosa*	Broad centurion (soldier fly)
Stratiomyidae	*Hermetia illucens*	Black soldier fly
Syrphidae	*Cheilosia variabilis*	Figwort cheilosia (hoverfly)
Syrphidae	*Chrysotoxum bicinctum*	Two-banded spearhorn (hoverfly)
Syrphidae	*Episyrphus balteatus*	Marmalade hoverfly
Syrphidae	*Xylota florum*	A hoverfly
Syrphidae	*Eristalis tenax*	Common drone fly
Tabanidae	*Haematopota pluvialis*	Notch-horned cleg fly
Tachinidae	*Exorista rustica*	A tachinid fly
Tephritidae	*Tephritis marginata*	A tephritid fly
Tetranychidae	*Tetranychus urticae*	Red spider mite
Tettigoniidae	*Leptophyes punctatissima*	Speckled bush-cricket
Therevidae	*Thereva nobilitata*	Common stiletto fly
Theridiidae	*Anelosimus vittatus*	A cobweb spider
Thomisidae	*Misumena vatia*	Goldenrod crab spider
Tortricidae	*Hedya pruniana*	Plum tortrix moth

*From yellow iris by the lake.

Survey conditions: 5 June 2023, Sunny/Cloudy, 13–21°C , ST 86586 30813 – ST 86618 30825.

CHAPTER FOUR

A Bustle in Your Hedgerow

Birds, Hedgehogs and Tree Frogs

*Watching the line of the hedge, about every two
minutes, either near at hand or yonder a bird
darts out just at the level of the grass, hovers a
second on labouring wings, and returns as swiftly
to the cover. Sometimes it is a flycatcher, sometimes
a greenfinch, or chaffinch, now and then a robin,
in one place a shrike, perhaps another is a redstart.
They are fly-fishing all of them, seizing insects
from the sorrel tips and grass, as the kingfisher
takes a roach from the water.*

Richard Jefferies, *The Life of the Fields*

We have touched on the crucial relationship between hedgerows and birds in previous chapters, but let's now look at the subject more closely, along with the hedgerow's crucial importance to our modern mammal and amphibian populations.

When I stand at the end of Jonathan's hedge at Underhill and look back at it from end-on, it reminds me uncannily of a huge old-fashioned A-frame tent such as Boy Scouts used to use. Or maybe an enormous Toblerone. But I prefer the tent image, because it closely echoes the shape as well as the function of our relaid hedge (the precise geometric shape is a 'triangular prism', if you must know). Just like a canvas A-frame tent, with its high ridge line and generously broad skirts, our hedge provides an excellent shelter and habitat for all the creatures within that call it home, as the rain, frost and snow cannot penetrate its bristling fastness and the ever-lively winds of South West England tend to whoosh up and over it rather than straight through.

Hedgerows today make superb habitats for any bird or animal that has inhabited these islands during the past ten thousand years or so in a landscape of woodland, grassland and scrub thickets. As we have emphasised before, and at risk of repeating ourselves, the well-managed hedgerow is a perfect substitute for the wild, thorny fortress of bramble and crab apple, hawthorn and blackthorn thicket which once dotted Britain's Holocene landscape.

A tight, dense hedgerow is the perfect nesting spot for multiple small songbirds, impenetrable to every predator from magpie to sparrowhawk. But birds also need hedgerows for their fruit, so the best cutting regime is always going to be a mix of tightly cut hedgerows and loose, sprawling hedgerows, festooned with spring blossoms, and autumn fruits and seeds. Birds also favour hedgerows of different sizes: generally speaking, the shy and modest bullfinches will nest in lower hedges some three to seven feet high, while the more flamboyant goldfinches like to be as high as possible, in hedges thirteen to thirty-three feet high.[1]

The ideal hedgerow though, skilfully and sensitively relaid at intervals of ten or twenty years, with its attendant grassy and weedy margins, can provide habitat and food for everything from the linnet that eat grass seeds, to cuckoos that eat caterpillars, especially hairy caterpillars, inedible to many birds but which the cuckoo is specially adapted to cope with.

In fact, the presence, or absence, of cuckoos, tells us a lot. A cuckoo in the summer will eat at least 150 hairy caterpillars a day, or some 22,500 per pair per breeding season, over the 75 days or so from mid-April to the end of June.[2] But birds have minimum viable population levels, just like any other species, so any landscape will require some 22.5 million hairy caterpillars to have an effective breeding population of cuckoos.

The last cuckoo strongholds in England are places like Dartmoor and the New Forest: areas specifically set aside for nature and very lightly farmed, if at all. Most of our agri-industrial landscape is not really 'countryside' at all, in the traditional sense, hence the disappearance of the cuckoo.

The cuckoo's favourite caterpillars include the drinker moth caterpillars, which need fen and marshy grassland; oak eggar moths, which like sallow and blackthorn; and garden tiger moths, which need nettle, dock and burdock. The garden tiger, incidentally, is down 90 per cent since 1956.[3]

However, look at hedges: or rather, look at the Underhill hedge, a model for how *all* hedges need to look.

Blackthorn for oak eggar moths? Tick.

Nettle? Tick.

Marshy grassland: it's pretty damp on the north side, but it would be better still if there were a good, boggy ditch alongside, when you might get drinker moths as well. And I'd be very surprised if we didn't have oak eggar moths on the land here too, though we haven't yet proved this with a survey. All good stuff for any arriving cuckoo, though.

In other words, yet again, we see how a hedge can comprise an astonishingly wide variety of accommodating habitats in a very narrow band of space.

There will be hawthorn berries for fieldfares, earthworms for blackbirds, thistles and dandelions for goldfinches, ash keys for bullfinches, snails for thrushes, mice and voles for owls.

Haw finches can even crack the stone *inside* the haw: 'a feat most people are unable to achieve without a subsequent visit to the dentist.'[4]

Open-field species like yellowhammer and whitethroat will be more often seen in arable land, feeding more among beans and peas than wheat – although pesticides and insect declines have greatly impacted their numbers.

The yellowhammer feeds in the field and verge, makes its nests somewhat low down in the densest hedges, but wants a tall hedgerow tree to sing from, to establish territory and attract a mate. You can easily see from this how the majority of so-called hedges that we see in today's countryside, lacking any wind-proof density, any standard trees, any substantial verge *or* any substantial depth or growth at the bottom, are utterly useless to the yellowhammer on at least four different counts.

Yet it's a most beautiful, colourful resident of our country. In Welsh it's called *Melyn yr Eithin*, the 'yellow bird of the gorse'. And among other things, what did the brilliant minds of the European Union target specially for removal from British fields? Gorse. (This is not intended as an 'anti-EU' point, merely an illustration of the rotten eco-sense of remote desk-bound bureaucracies, of any affiliation.)

The Scandalous Dunnock

Herbicides have also greatly reduced numbers of so-called 'weeds', essential for such humble little birds as that supremely hedge-dwelling bird, the dunnock, which loves grass seed, chickweed and plantain seeds fallen to the ground. You wouldn't have thought this simple diet would be hard to find in a flourishing countryside – but the dunnock is now on the Amber List of threatened species.

The name 'dunnock' is a charming one, by the way. The '-oc' or '-ock' suffix simply denotes the Old English for 'little' (as in buttock, and, er, bollock) and dun means a slightly dingy greyish-brown, of course. So, we know that our ancestors right back to Anglo-Saxon times called the dunnock 'the little brown one', while the robin they often called the ruddock, or 'the little red one'. How sweet.

The dunnock should more correctly be called a hedge accentor, according to some modern ornithologists, but I suspect the old folk name of dunnock will survive a lot longer.

Other old names for the dunnock are the hedge sparrow, the hedge betty or the sugge, as in Sugnall in Staffordshire and Sugworth in Devon. They mean 'wood with the sparrows or dunnocks' and 'enclosure with the sparrows or dunnocks' respectively. An enclosure strongly suggests a space surrounded by hedges – so another glimpse of our bird-rich past.

The OED has examples of the word sugge from the early 1400s, saying that this is a shortened version of the earlier haysugge, hay again here denoting hedge. But what does sugge *mean?* It is, alas, not clear, nor does it seem to be related to any other similar word. But either way, sparrows and dunnocks obviously have a long association with our hedges. Not surprising, then, that our hedges and our sparrow and dunnock population have been declining in tandem in recent years.

The dunnock lives in hedges and feeds quietly and modestly on the ground for the most part, so was often used in old sermons as a symbol of humility, not to mention chastity, in the belief that they mated for life and gave us a sobering lesson in the joys of conjugal fidelity.

Modern field studies have revealed that this is utterly untrue and that the dunnock in the humble hedgerow has perhaps the most polymorphously perverse sex life of any British bird, with wife swapping, threesomes, group sex and even sexual congress with threats being the norm.

For some reason, a pair of dunnocks will often tolerate another unpaired beta male in their territory. In most other species, the alpha male would simply drive the beta away or even kill it. The presence of the beta is a headache for the alpha dunnock, though, as he now has to keep guard over his good lady wife the whole time – and the wife seems extremely keen to get away from him and to mate with the beta, which she usually does.

Those old sermons should really have used the dunnock, as an example of the most abominable sinfulness and depravity, taking as their text, perhaps, the lines from Ecclesiasticus, 25:19: 'All wickedness is but little to the wickedness of a woman.'

There's an even darker side to the dunnock. The female may actually mate with the beta male to dissuade him from destroying her eggs in the nest: in other words, she consents to sex in exchange for his not killing her offspring. And then when the brood is hatched, they will receive the food brought by two males, not one.

Dunnocks mate more fervently than any other small bird, an impressively vigorous once or twice an hour for an exhausting ten days, although at least, as ornithologist Mark Cocker says, 'copulation itself is remarkably brief'. The male penetrates the female in a split second and then does a peculiar jump over her, so that for a long time observers of this behaviour didn't realise that mating had taken place at all. Perhaps they thought it was some kind of jolly game of leapfrog. Given that there are still about three million pairs of dunnock in the UK, during mating season, in any one hour there may be as many as six million 'games of leapfrog' occurring in our hedgerows. Quite a thought.

Diversity of Hedgerows = Diversity of Birds

For the greatest overall health of our bird populations – and thus of our entire ecosystem, since they are very much the canary in the coalmine – the greatest possible diversity of hedgerows and hedge-cutting regimes is best. While most of our birds are best suited to the tall, dense, thorny thicket of a hedge, achieved by conservation hedge-laying and very sensitive cutting, some will prefer a slightly tighter or lower hedge.

The 'tall, lush, thickety' group would include everything from robins and song thrushes to willow warblers and nightingales, while the yellowhammer and linnet prefer shorter, neater hedges.

Dr Max Hooper found that while stump-line remnant hedges are unsurprisingly the worst habitat, those recently laid take a while to get back full numbers, but when they finally produce that much sought-after blossom-and-fruit-laden thickety structure, 'the clear winner was a very overgrown hedge that had begun to invade the fields with escaped blackthorn. ... For birds, bigger really is better.'[5]

That miserable scarred and flailed stump-line, however, that is the remnant hedge, may be the current condition of as much as 50 per cent of our remaining 400,000 kilometres of hedgerow – and far more in certain landscapes, for instance arable lands. This is an underappreciated disaster – and an ongoing one. Relict hedges urgently need restoring, far more so than new ones need planting.

From their surveys of a hawthorn hedge, Pollard, Hooper and Moore established that a 'remnant' or 'relict' hedge supported all of 4.4 breeding pairs of birds per 1,000 yards on average, while at the other end of the scale, the richest was the hedge 'with outgrowths', meaning it was allowed to spread here and there into the neighbouring land. This happened especially often with blackthorn, creating bays and scallops, but hawthorn hedges were the richest for birds; a dense and vigorous hawthorn hedge supporting 34 pairs of nesting birds per 1,000 yards, across 19 different species.

Thirty-four, against less than five in a relict hedge. You can see why dying hedgerows are such a disaster: they effectively cause a fall in numbers of hedge-nesting birds of some *87 per cent.*

The open-field system of medieval times, as we know from that extraordinary surviving example the village of Laxton in Nottinghamshire, had no hedges at all – though no doubt plenty of nearby areas of coppice, scrub and woodland – and supported millions of skylarks, partridge both grey and red-legged, and lapwing, perhaps the most familiar birds of all in the day-to-day life of the medieval field worker or Piers Plowman. Unsurprisingly, the plump partridge was also a common dinner staple.

But all the other birds of more recent well-hedged countryside, like blackbirds and blue tits, though absent from the medieval open fields, would have inhabited the numerous coppices and woodlands, especially woodland edges instead. They would not have been rarities.

Lost Birds

Two summer visitors that really love our hedgerows are the common whitethroat and lesser whitethroat – though, ironically, it's the common whitethroat that is now on the Amber List. The UK had a population of some five million pairs up until the 1960s, but numbers have crashed with only partial recoveries and it is much rarer now.

Droughts in the Sahel, its winter destination, are partly to blame, as well as human population growth in Africa and consequent overgrazing. Loss of vegetation then leads to desertification and lower rainfall still.

This is a disaster for many species of summer visitors to Britain, which then return to sub-Saharan Africa for winter: as well as the whitethroats, this includes pied and spotted flycatchers, nightingales, swallows, turtle doves and more. We can do what we can in Britain, but other countries are out of our hands.

The blackcap, the 'northern nightingale' with its lovely song, is another quintessential bird of hedgerow and thicket. It is also a migrant southwards in winter, but now more and more birds are staying put and surviving here through the coldest months: all the more reason to create or leave well alone any thick, weatherproof hedges in late autumn for them. The thickest and wildest hedges of all may give us whitethroats, but then where young trees burst up through the scrub, you may get grasshopper warblers, while these 'dense thorn castles', in Benedict Macdonald's evocative phrase, admitting little light, are the favoured haunt of nightingales.

Later in the year, in addition, the huge abundance of ready foods, nourishment, energy and calories on uncut thickets and hedgerows is crucial to half-starved migratory birds arriving in winter, like redwings and fieldfares, bramblings and goldcrests.

In fact, the higher the bird numbers, the more good hedgerows matter. And they are falling precipitously, along with much else. In the 2019 State of Nature report, of the 8,431 species of all kinds surveyed in Britain alone, 1,188, or 15 per cent, are threatened

'Jonathan's hedge'. The famous hedge at Underhill just six years after being relaid, and now definitely a primeval thicket. Shown here with Jonathan.

Classic scrub pasture naturally recreated at Knepp by the cattle and already looking very like prehistoric Britain: the equivalence, and importance, of hedgerows in farmland today is very clear. Photo courtesy of Richard Hoare.

Aberdeenshire hedgerow. Neatly cut, no doubt annually for traffic safety, this is a good, tight hedgerow offering excellent wildlife habitat. Not all flailing is bad news. 'Stracathro Hedgerow' by Andrew Wood.

A wonderful thickety hedgerow along with promising-looking grassy verge, Lincolnshire. 'Hedgerow and Field' by J Thomas.

The first redwing seen at Underhill – a great sign! One of many feeding on winter berries still on the hedgerows on 16 December 2022. Farmland hedgerows flailed hard before this would have left little fruit.

Another hedgerow fruit-loving winter visitor to Underhill: the fieldfare.

The fantastic 'nightingale hedge' at Knepp at dusk – just when they start to sing. Photo courtesy of Richard Hoare.

Staffordshire hedgerow in winter. Threadbare and exposed, a typical degraded hedgerow today. It would offer much better protection if it could be relaid again. 'Winter Hedgerow' by Philip Jeffrey.

Relict hedgerow, South Oxfordshire. A good strong hedge but hopelessly open to weather and predators. This could easily be relaid or coppiced at ground level, with great benefit. 'Hedgerow by the Footpath' by Steve Daniels.

Another view of the Underhill hedge with a fallen oak. A sad loss of a magnificent tree, but it became a haven for a whole new range of wildlife and will remain so for a century or more.

Spring flowers at Underhill: bluebells, orchids and more.

An end-on view of the Underhill hedge, showing the dense, desirable pyramidal shape, which offers all different heights and zones.

Home-made hedgerow pie. Tastes better than it looks.

And how it started …

The European tree frog. And we want them back.

Does all hedge-laying have to produce a hedge? Some self-grown blackthorn shrubs on our land that I cut and laid into an instant thicket. Very satisfying.

Hedge-laying: ancient …

… and modern. Both can be equally effective with the right skills.

A stretch of beautiful Salisbury Plain, unfenced, grazed, with scattered thickets: scrubland.

An ideal mature hedgerow: festooned with blossom, with mature trees, long grass and short grass, and still plenty of room to grow crops.

with extinction and 2 per cent are already extinct.[6] That's around 150 species: gone forever. Farmland birds have fallen 54 per cent since 1970 and woodland birds another 25 per cent.[7]

Five woodland specialists (lesser spotted woodpecker, lesser redpoll, spotted flycatcher, capercaillie and willow tit) have declined by over 80 per cent relative to 1970 levels, with the latter down by 94 per cent.[8]

Now rare cirl buntings were common even in Wembley Park and Wimbledon Common in the late nineteenth century. Today, just five hundred pairs or so are confined to South Devon, with a few more reintroduced in Cornwall. They require dense hedgerows to nest and grain to feed upon – but hedgerows in our modern grainlands tend to be the most ruthlessly cut, flailed and minimised of all.

No wryneck has returned to England to breed since the 1970s – a remarkable, unique bird of the woodpecker family. Yet in the mid-nineteenth century, William Yarrell said that they were so common that they were often kept by country children; and in 1909 they were described as 'plentiful and generally distributed' throughout Kent, according to *A History of the Birds of Kent*.[9]

Their Latin name, *Jynx torquilla*, is a strange and exotic one: it seems wrynecks were anciently used in some obscure ritual in Greece to hex, spellbind or 'jinx' a lover, hence the first part; and *torquilla* means little twister, as in snakey, or snake bird, from the wryneck's weaving of its head and neck back and forth like a snake. People obviously found it an uncanny and fascinating creature – and knew it intimately.

And as we mentioned in the previous chapter on insects, a lack of certain bird species can also suggest how many plants and/or insects we have lost. A breeding pair of wrynecks requires 800,000 ant pupae over one summer, for two broods. In Benedict Macdonald's mind-boggling calculation, even 'a small viable population of 50 wryneck pairs would then, in the course of one summer, require close to 50 million ant pupae to survive. That's a landscape with up to 3.4 million anthills.'[10]

And the landscape that produces such countless numbers of anthills is the herbivore-grazed wood-pasture mosaic of yester-year, Britain's equivalent of the Serengeti, the wryneck's natural home. In fact, it is safe to say as a general rule that pretty much *all* of our native birds and animals, and our trees and flowers too, are best suited to a colourful jigsaw puzzle of majestic single trees, copses and thickets, and wide, open grazing lands: strangely akin to parkland, yet created without human intervention and long before the advent of farming.

Macdonald notes that the same can be said for our eighteen native species of bat, all evolved for just such a patchwork or jigsaw terrain: not a monolithic agri-zone, but a rich mix of flower meadow, thorn stands and ancient, gnarly trees full of rot holes and hollows, a world absolutely abounding in insects, especially the bats' juicy favourites, beetles and moths.

This is also true of such overlooked families as lichens, mosses and liverworts, which again favour marginal edge habitats rather than dense forests. Lightly grazed and unimproved grassland enclosed with frequent densely grown hedges, dotted with trees both mature, ancient and dead, is an excellent substitute.

The natural mosaic of ancient Britain meant woodlands and grassland in perpetual competition with each other, vast herds of grazing animals constantly knocking back the trees, but then their relentless pursuers, the carnivores, driving them onward in the ecology of fear, unwitting guardians of the trees.

Returning to wrynecks and their gargantuan appetite for ants, and turning Macdonald's illuminating back-of-the-envelope calculations on their head, you get a vivid sense of a landscape we have never seen and find hard to picture: if in 1909 wrynecks were so common in Kent as to be familiar childhood pets, imagine the numbers of anthills there must have been even in the small-scale, well-farmed Kentish countryside. It was teeming with wildlife in a way that we can only imagine when my own grandfather was a boy: born 1900, died 1986. Not so long ago.

Hedges cannot recreate every aspect of wild wood pasture, of course, nor bring back every bird. In Spain, even cranes will feed in wood pasture, or *dehesa*, as such ancient and undrained wood pasture very often has enough bogs for them to feel at home.[11] They may well have done so here in the past, though they were virtually extinct as breeding birds by the reign of Elizabeth I, possibly by 1600. Today, a tiny number are breeding again in Norfolk, though to think that they were once perfectly common in all the crane-based place names of Britain, now either urbanised or heavily farmed, tells you how wild and wood-pastured such places must have been only a few centuries back: places like Cranmere and Cranleigh, both in Surrey, and Cranborne in Dorset.

It sometimes breaks the heart to picture how things were, how wild and magnificent: to picture the crane's gigantic, awe-inspiring silhouette against the medieval sun, its seven-foot wingspan, a feathered Concorde of the skies, casting its giant shadow over today's suburban Surrey. Now we've got Gatwick Airport nearby instead.

Hedge of Life

A well-cared-for veteran hedgerow in the English countryside might be home to at least a dozen species of our native mammals. This is out of a total of about forty-nine native species, though this is becoming increasingly hard to define. Is the beaver now a native species or not? Nevertheless, as a safe haven, feeding and breeding ground for some 25 per cent of our mammals, it's another measure of just how vital hedges are to a healthy ecosystem.

One animal in particular is emblematic of our hedgerows: the hedgehog. Also called the hedge pig, or in Danish the *pindsvin* meaning 'stick pig'. There's something about these small and rounded nocturnal little creatures that is irresistibly cute. No wonder it's Britain's favourite mammal, according to a 2016 survey, winning a whopping 35.9 per cent of the vote.[12]

Hedges are crucial to hedgehogs, but, intriguingly, study after study has shown that they make up only a small fraction of a typical hedgehog's range. They also roam surprisingly far and wide, snuffling about and foraging in nearby fields and woods. Yet regarding the hedgerow, 'that tiny fraction is where the hedgehogs opt to spend most of their time'.[13]

There are many reasons for this, one of the less appreciated being that the hedgerow offers both central heating and air conditioning, as required. One study recorded the average temperature at night as 10.7°C in the middle of arable fields, but 11.4°C in hedgerows: a crucial difference for many smaller mammals, including the hedgehog. Along with the welcome warmth, there's food, ample bedding for nests in the form of fallen leaves and dry grasses, a place to hide, thorny and thistly protection from larger predators, and a useful aid to navigation in the hedgerow itself.

A hedge is intended as stock barrier, a boundary marker, a kind of living, growing fence. But if you wanted to create a habitat very specifically for our native hedgehogs, it would be hard to devise anything better.

With all this in mind, you would expect us to take especial care to preserve our hedgerows, given that these are the preferred habitat of our favourite mammal – yes?

Alas, and unsurprisingly, the decline of the hedgerow has meant a decline in hedgehogs as well. In the 1950s, there were still some 36 million hedgehogs in the UK, but this number has fallen catastrophically to half a million, and is falling still. Our favourite mammal is going the way of the large copper butterfly, the wryneck and the brilliant green tree frogs properly native to Britain. Our country is far more denatured than most of us even realise, with such lost amphibians now erased from any popular memory and hedgehogs facing the same fate. Our environment could yet become still more dramatically simplified, flattened out and monotonous in the next few decades – and such environments are, as we are tirelessly warned, always extremely unstable.

Hedgerow quality is also essential – a major problem for our beleaguered hedgehogs. A 2007 Countryside Survey reckoned only 48 per cent of our hedgerows were in structurally good shape, but this must be closer to 30 per cent today, if, as Tom Moorhouse points out, 'good condition' includes having a flourishing grassy verge, full of wildflowers, since this such is a vital element for wandering, foraging denizens of the hedgerow.

The result of these lacks and losses can be terrible. Moorhouse cites the findings of one of his research students, Carly, who did a hedgehog survey with footprint tunnels in the spring of 2012 over very large areas of Oxfordshire farmland, 6,000 hectares in all. And she found... absolutely nothing. Mile after mile of silent Oxfordshire farmland 'yielded not one set of hedgehog prints'.

Both experiment and anecdote have confirmed that hedgehogs (and urban foxes), given the choice, will frequently return to the urban environment of gardens and parks rather than to the bleak, lifeless wastelands of the agro-industrial landscape, where food and shelter are so scarce.

Hedges are also crucial fox dens, with foxes being another of the twelve or so mammals that make their home here. And their decline echoes that of the hedgehog. With habitat and natural prey animals much reduced, many a fox has upped and moved into the city, where, with little to hunt, except perhaps the occasional rat or mouse, he becomes a scavenger, not a killer. And when brought out into the countryside – often put there by well-meaning 'animal activists' – the overweight, under-nourished, semi-domesticated canine sits and stares dumbly around, utterly incapable of feeding himself by hunting, accustomed as he is to off-scourings of fast-food outlets and waste bins. Perhaps this once beautiful wild creature has become something of a mirror image of ourselves with an unfortunate taste for late-night doner kebabs. Both we and the fox need rewilding.

Other small mammals that love well-managed hedgerows are the mice and voles, including bank voles, harvest mice and hazel dormice. Hedgerows that connect to woodlands, no matter how small in acreage, enrich their mammal population, and vice versa, offering a far greater range of vegetation, and hence foraging opportunities, than just an isolated copse or a disconnected hedgerow.

If you want to know what's eating your rosehips in autumn and winter, examine the cut just below a missing spray: if there is a neat diagonal cut, then you can assume the hips have been scurried away as a juicy and highly nutritious prize by either a field mouse or harvest mouse or a bank vole. Mice will discard the flesh, surprisingly, and eat the oil-rich seeds, while bank voles eat the flesh, or some of it. So, the wildlife Sherlock Holmes, if they find any rosehip remains on the ground beneath the cut spray, might even be able to determine whether they've been left by a mouse or vole.

By far the commonest hedgerow mammals are the least seen and least known by the general public: the wood mouse, yellow-necked mouse, bank vole and common shrew made up 97 per cent of all hedgerow animals found in one survey on arable land.[14]

The lovely little harvest mouse, too, used to be common in our arable fields – hence it also being known as the field mouse in older texts – but is now found only in hedgerows, once again a last redoubt for a once familiar and much-loved creature. Especially so in diverse, mature hedges adjacent to ditches and/or unimproved pasture. On our land, the brilliant roving wildlife expert, Gareth, found a harvest mouse nest in the reeds, between rough grass and the river. They're incredibly small and hard to find, but thrilling if you do: a neat little structure about the size of a golf ball.

All of these busy little rodents – harvest mice, bank voles, shrews and others – originally evolved to inhabit the evolutionary niche of the sunny, fruit-some and nut-some woodland edge – hence their love of the hedge, a high-calorie buffet bar with added temperature control. A crucial consideration for any tiny but warm-blooded creature: the common shrew's heart, for example, runs at a somewhat manic 1,000 beats per minute. Shrews are primarily insectivores, having barely evolved in 50 million years or so. If you want to know what your direct ancestors looked like soon after the demise of the dinosaurs, look at a shrew.

And, of course, the hedgerow also offers an inexhaustible richness of invertebrate life, as we have seen. They eat three quarters of their body weight each day, and one and a half times that for a pregnant female. (Although this still can't compete with the Polyphemus moth caterpillar, which eats 86,000 times its own body weight in two months. I'd have to eat 8,256 tonnes of food in that time to match it.) Interestingly, the previously mentioned survey on arable land also found numbers of water shrews, showing how hedgerows can create an invaluable damp and boggy habitat too.

A Long-Lost Frog

Which brings us to a little creature that I find strangely haunting, a small, bright green symbol of what we have lost: our native, but extinct, tree frog.

Nearly all our remaining amphibians and reptiles can be found at times in hedgerows, especially if they have a ditch or a pond nearby. But at least one of our amphibians, which *ought* to be native, is now missing. And this has happened startlingly recently, as is clear from this remarkable passage:

> The little Frogge of an excellent Parrat-green, that usually sits on trees and bushes, and is therefore called *Ranunculus viridis*, or *Arboreus*; but hereby I understand the aquatile or water Frogge whereof in ditches and standing plashes, wee may behold many millions every Spring in England.

Thus wrote the wonderful Sir Thomas Browne, in his book published in 1646, *Pseudodoxia Epidemica, or Vulgar Errors*.[15] And he mentions our lost tree frog as if it were a perfectly common and unremarkable animal to see in the trees and bushes of the English countryside.

Sir Thomas Browne (1605–1682) was a Norfolk doctor and writer, and admittedly Norfolk was still famous then for its many fens and frog-friendly places, even after the ditching and draining to create hugely productive farmland had already been going on some centuries. Browne wrote on a fantastic variety of subjects, anything that interested him, his most famous works being *Religio Medici* (A Doctor's Faith) and *Hydriotaphia, Urn Burial*, all composed in a beautifully ornate seventeenth-century prose style, which is still a rich pleasure today.

But what are we to make of his allusion to tree frogs, and how does this relate to today's hedgerows? Especially since he writes of these frogs numbering 'many millions'.

Even allowing for some exaggeration, what a haunting sense of lost abundance. Derek Gow has also pointed out that tree frogs are 'very very noisy. If you're looking at something which attracts the attention of someone 400 years ago, it's because they're looking at the noisiest frog in Europe.'[16]

Among all the more high-profile and laudable initiatives to rewild our countryside with charismatic animals like the beaver and the lynx, what about the tree frog? And then there are the reptiles: our island once had twelve reptile species, but we are now down to just five – and they are *all* at risk. Among the amphibians, the European pond terrapin, the Aesculapian snake, even the fire salamander were all once native. Today, the natterjack toad and the smooth snake are now officially Endangered, according to the Amphibian and Reptile Conservation Report of 2021. Even the common toad is classed as Near Threatened.

One way to protect these amazingly ancient and charismatic creatures, as we know, is to create the right environment – and the wet, dense, humid hedgerow could be ideal. Our lost tree frog, Sir Thomas Browne's frog, in its European strongholds, is a cheerfully accommodating creature, equally happy in mixed woodlands, gardens, vineyards and orchards, as well as beside any bodies of water, large or small. Spawning can take place in something as small as a ditch or puddle, hence the suitability of a hedgerow with a ditch as an ideal habitat.

While we're on the subject of lost frogs, there are two more: the agile frog, *Rana dalmatina*, and the moor frog, *Rana arvalis*, both lived in England until at least Roman times. There was also a small colony of rare pool frogs, *Rana lessonae*, in Norfolk, thought to be an introduced species but finally appreciated as native in 1999 – the year the last one died.

It makes one imagine a much wetter, swampier and less drained Britain as being a paradise for frogs, but a more generally wild and thicket-grown landscape would be just as frog-friendly. The common tree frog doesn't favour dense, dark, lichen-grown old woodlands as much as sunny open scrub with plenty of bramble and gorse. Or a well-hedged countryside with lots of ditches and ponds.

A less popular resident of the hedgerow is the grey squirrel, blamed not only for eating all the hazelnuts but for inhibiting new hazel growth as well. A glance over our own, squirrel-rich wild patch suggests this is untrue: we have many new hazel bushes growing on our seven acres, mostly near to existing hazel trees, and many I suspect actually planted by squirrels for their winter cache, and then forgotten or simply not needed. I always find it endearingly optimistic of a squirrel to believe when he buries his cache that he alone will be the one to return to enjoy it in the colder months. Any passing squirrel can smell a stash of hazelnuts below the soil and gobble them up. But I suppose it all works out overall, in some kind of anarcho-communistic way. And I have stopped worrying about the 'damage' done by this non-native species. I'm not seeing any real evidence at first hand.

There is one more animal that loves the darkness, shelter and privacy of the dense old hedge. The mole.

There will be many molehills, generally unseen, inside bushy hedgerows, because where there's leaf mulch there's worms; and where there's worms, there's moles. They have to eat every few hours, and in freezing winter conditions, when they're unable to dig through the frozen surface of open fields, the hedgerow again will be warmer with easier digging. Like squirrels, moles too will keep a little store of food laid in for hard times, although in moles' case it *is* a bit horror film: they bite the heads off any surplus worms and store them alive in a special tunnel. Meanwhile, the said worms sit quietly in the dark, quickly trying to regrow their heads so they can make good their escape before their hungry captor returns.

The truth is that so crucial is the thickety native hedge that almost *any* of our birds and animals could be included in this chapter. Pheasants also love hedges. Deer will disappear into larger ones and double-row hedges may accommodate a muntjac for much of its life. Bats use them for navigation, badgers build their setts in their ancient banks, amphibians, reptiles and some butterflies, like the peacock and tortoiseshell, will hibernate in them (another reason to treat your hedges sensitively between December and February), even producing their own glycerol, which acts like antifreeze to see them through the cold.

The hedgerow can still help to save our birds, our animals, even our amphibians and reptiles – not alone, but it can contribute enormously, since it is such a key, irreplaceable feature of our mosaic-substitute farmed landscape. A hedgerow is the beautiful green and red, white and gold stitching that holds this gorgeous mosaic together. If the stitching becomes frayed, the mosaic begins to fall apart…

It is clear what needs to be done, and just how great the rewards for all would be.

Hedgerows versus the Wasteland. That's my motto.

CHAPTER FIVE

Hedges

The Living Food Bank That
Keeps On Giving

Here the industrious huswives wend their way
Pulling the brittle branches carefull down
And hawking loads of berrys to the town
Wi unpretending skill yet half divine
To press and make their eldern berry wine
That bottld up becomes a rousing charm
To kindle winters icy bosom warm
That wi its merry partner nut brown beer
Makes up the peasants Christmass keeping cheer

John Clare, *The Shepherd's Calendar*

The hedge is an outdoor salad bar that continually replenishes itself, while giving us wild garlic, garlic mustard, young nettles, hawthorn leaves, sorrel, corn salad, ground elder, sow thistle, lady's smock, chickweed, fat hen, cow parsley and dandelions – both flowers and leaves.

And then later in the year, it magically transforms into a fruit stall: rosehips, haws, blackberries, crab apples, sloes, rowanberries, wild redcurrants, blackcurrants, raspberries, strawberries... I have personally picked, cooked and eaten all of these hedgerow fruit and more.

And not forgetting hazelnuts – if the squirrels don't get there first.

An intelligently timed cutting and laying regime can produce a hedgerow absolutely bursting with all this continuous food for free in practically every season, and year after year, in perpetuity.

Country people having been using the hedge for centuries as a source of free food. Indeed, when we pick blackberries we are doing exactly what our ancestors ten thousand years ago would have done, when the ice sheets had retreated from Britain and the landscape reverted to forest and wood pasture, dotted with thickets full of brambles.

But today, people are sadly distanced from any notion of finding food for free, while global supply chains are beginning to buckle, and when an additional crisis hits – Covid, for instance – we see the shelves in our supermarkets empty out with alarming speed.

Much of the ultra-processed and plastic-packaged imitation food sold to us is so low in nutrients, but so high in calories, that it's a blindingly obvious cause of our appalling rates of obesity. You don't have to be a scientist to understand why. Although taking in far more calories in a day than they need, the modern food consumer will still be nutritionally starved, their bodies crying out for healthy fats and aminos, calcium, magnesium, zinc – and so they will keep on eating.

How can hedgerows help? Because they are a living food bank that keeps on giving, and have been used as such for hundreds, probably thousands of years. And they provide just those wild foods that are the polar opposite of junk food: low in calories, but off the scale in nutritional value.

We will take it as read that the wild foods under consideration in this chapter come from hedges that haven't been sprayed with the usual arsenal of agrotoxic chemicals. Hedgerows in pastureland, in remote uplands, by the coast, on organic farms or on private land managed for wildlife are probably your best source. And don't forget school playing fields and public parks either. I've collected sweet chestnuts off Hampstead Heath, roasting them on an open fire fuelled by clean offcut wood taken from a skip, almost certainly in contravention of any number of rules, I'm delighted to say – and they were delicious.

Salads and Greens

Let's start by looking at all the wild foods and remedies available just in the hedge at Underhill, using the list of plants compiled by Jenny Bennett back in chapter two.

After blackberries, the green leafy abundance of the hedgerow in March is probably what most people are familiar with. And the choice on offer at this particular salad bar can be dazzling. The Underhill hedge alone, for instance, offers cleavers, ramsons/wild garlic, cow parsley, dandelion and wood dock.

Wild garlic, like all spring greens, is lovely when quickly fried in butter, but also makes a fantastic soup if you mix it up with equal quantities, more or less, of nettle tops and some good bone stock. And the absolutely best thing you can do with a leaf or two of wild garlic, and/or dandelion, is to lay them directly inside your Cheddar cheese sandwich when you're out for a long springtime walk.

Cleavers or goosegrass we have mentioned already: the best and most traditional thing you can do with this abundant plant of many a hedgerow is, as the name suggests, feed it to a goose, and then eat the goose. But you can also fry the young shoots lightly and eat it direct.

And yes, all species of dock (*Rumex*) are edible – in moderation.

As a general rule, *all* wild foods should only be consumed in moderation. Our ancestors would have foraged from a vast variety of plant foods and, even today, Kalahari bushmen have been found to regularly eat over one hundred different kinds. So, with dock, for instance, you shouldn't eat too much because it contains quite high levels of oxalates, which can cause kidney stones. But so do rhubarb, spinach and almonds, and no one says they're bad for you. Even broccoli contains bad stuff like thiocyanates. If you were to eat five kilogrammes of broccoli at a sitting, you would probably die. But people tend not to anyway.

In general, a little of everything is best. One of the few food groups that contains no anti-nutrients is meat – because an animal's favourite ways of not being eaten are to hide, fight back or run away. Plants can't hide or run away, so they fight back with anti-nutrients – except fruit, of course, which *want* to be eaten, so you can spread their seeds around the country-side when you poo.

It's all very simple really.

Back to your hedgerow salad.

You could also throw in some hawthorn leaves and some petals from the field rose and the dog rose, to be really cheffy, to create a stunning salad from just this one stretch of hedge.

One other plant on Jenny Bennet's list from the Underhill hedge is soft rush, also said by some to be edible; apparently it is well known in Chinese herbal medicine, while the Japanese make it into a tea, called *hui sup*. But I haven't found much solid evidence for its edibility, let alone deliciousness. Many wild foods fall into the category of 'may be edible – it's not exactly clear', and, if so, I advise caution.

Wild food does tend to taste strong, compared to the bland pap we are used to from the supermarket. Yet it is precisely those strong, slightly bitter flavours that signal its nutritional value. As the year progresses, the leaves' protective chemicals build up, and I wouldn't recommend most of them after May.

One fine exception to this, though, is nettles, if they grow on your own patch. Of course, they need cooking first: steamed, plain or sautéed in butter. But when May or June comes around, if you cut your nettle patch to the ground – and assuming you then get some good summer rain – within just another three weeks or so, you'll have a whole new crop of bright young nettle shoots to eat a second time around. I have eaten new-growth nettle tops as late as October, with no negative consequences. But if you try eating stringy older nettles after June, you'll find that: a) they taste bitter; and b) they are laxative.

You have been warned.

Comfrey leaves, by the way, only recently went off the menu, although many older wild-food books recommend them. They're now known to contain high quantities of rather toxic alkaloids. Pliny recommended adding them to stewed meat, Culpeper recommends them too and even John Lewis-Stempel eats comfrey fritters, reminding us along the way that comfrey 'has more protein in its leaf structure than any other British wild plant'.[1]

If you avoid the comfrey, though, you might still use his 'leaves in batter' recipe with something else, coating the leaves in duck-egg yolk, then frittering them with goose fat, chestnut flour and hazelnut oil. Sounds fantastic. What about with wild garlic and some very young horseradish leaves?

Hawthorn leaves are about the best tree-leaves to eat in spring (though you can also try beech and lime). They used to be called bread-and-cheese by country children, in honour of which Streeter and Richardson suggest laying a bunch of fresh young hawthorn leaves on pieces of fried bread, topping with cheese and grilling until bubbling.[2] Delicious.

There's a strong and unsubtle clue about the healthy powers of yarrow merely in the name: it's related to the Old English for 'healer', *gearwe*, while other folk names include woundwort and staunchwort. The Romans called it *herba militaris*, 'soldier's herb', for its healing powers.

It used to be used instead of hops to preserve and flavour ale, and I've done a 24-hour cold-tea brew of yarrow flowers and meadowsweet flowers, which was excellent. You can also do the Anglo-Saxon thing with your ale, adding either herb, or any other suitable one – flowers or leaves – to your glass before you pour on your wine or your ale. Use spicy herbs for wine, pungent ones for ale, and leave them for a few minutes to infuse.

The results are enjoyably unpredictable.

The truth is we are terribly dull nowadays with our foraging and cooking compared to our forebears. Consider a surviving recipe for an English omelette with herbs, from the early Middle Ages: it contained 16 eggs, chopped dittany, rue, tansy, mint, sage, marjoram, fennel, parsley, beets, violet leaves, spinach, lettuce and pounded ginger.

Who would mix ginger with eggs and mint nowadays? But it might be delicious. Violet leaves are edible; tansy and rue are often overlooked old English pot herbs, along with that old staple of the herb garden, lovage; and the dittany is now almost completely forgotten.

I actually take this to be what is more commonly called dittander, also erroneously dittany (*Lepidium latifolium*), another member of the cabbage/brassica family, so many of them with powerful health-giving properties and not the dittany *Dictamnus albus*, which doesn't appear to be edible.

Dittander is a perennial of damp ground near the coast, used as a spicy condiment before the horseradish was introduced from the Near East, and recommended by medievalists for leprosy, hence surely its continuing presence in the grounds of the Hospital of St James and St Mary Magdalen in Chichester, founded in the twelfth century.[3]

What about carbohydrates, often a sought-after rarity for the novice wild forager? Roots and tubers are one of the best and healthiest sources of carbs – healthier than grass grains like wheat and barley. Have you ever heard of anyone with a potato or sweet potato intolerance? It's very rare.

But what did the wild-food forager eat before the cultivation of the humble spud or the earliest spelt and emmer grains? John Lewis-Stempel maintains that the tubers of silverweed were the ones we ate in Britain before the arrival of the potato, which is persuasive, but I imagine that we ate dozens of different bulbs, roots and tubers in the very old days: silverweed, burdock, horseradish, wild garlic, bullrush, and no doubt many more that we've forgotten. Silverweed and bullrush, along with flowering rush, reedmace and other plants of the ditch and wetland, have roots that can be baked and then ground into flour.

But another absolute staple – and the one that appears at Underhill – is the pignut. You have to get really palaeolithic here with a pointy digging stick but, if you enjoy such things, it's fun. You dig straight down beside the pignut and then along a bit to find the walnut-sized tuber, which is always at a right angle to it. 'A sweet chestnut crossed with a radish' is one very good description I've heard. It's excellent eating, especially roasted, like all roots.

I should add that Culpeper warns us that pignut can be a little too good for you, though, and 'provoke lust exceedingly and stir up those sports she is mistress of'. Treat with caution.

Another root that you can find everywhere and eat with a clear conscience, as it's so common, is hedge garlic, or jack-by-the-hedge. It's a little like horseradish. Grate it raw over pretty much anything for added zing: roast potatoes or steamed greens, fresh salads, add a little oil; it's a terrific flavouring. And being a brassica of the cabbage family, it has enormous health benefits.

Cultivate eagerly all along your hedge.

My absolute favourite wild green has to be wild garlic, though. If you mix it 50/50 with young nettle tops, it's a fantastic side vegetable, but it also makes a terrific pesto. The original Italian version uses pine nuts and parmesan, of course, but you can make a delicious UK-based equivalent (in fact, it's probably cultural appropriation) if you mix a handful of raw wild garlic leaves, some roasted hazelnuts, any strong hard cheese and a good gloop of olive oil. Blitz it in a food processor and devour.

Autumn Fruit and Nuts

With its abundant blackthorns, the Underhill hedge also shouts sloe gin to anyone fond of that particular winter warmer. Birds do eat sloes as well, especially the thrush family, but it's rare to see a blackthorn stripped entirely, and I think there's enough to go round for all. Sloes are one of those typical off-the-scale wild fruits in terms of nutrients, antioxidants and the rest. You can tell from the colour alone: anything naturally that dark purple is going to do you good, as long as it's edible.

Raw sloes will give you some useful vitamin C (10 milligrams per 100 grams) and vitamin E (5 milligrams per 100 grams), but are more useful for high levels of potassium and magnesium (453 milligrams and 22 milligrams per 100 grams, respectively). However, it's the compounds like phenols and flavonoids that really boost you, with one study reporting that 'ethanol extract from blackthorn fruits could upregulate miR-126 and miR-146a expression levels and therefore downregulate the expression of the pro-inflammatory cytokine IL-6 and adhesion molecules ICAM-1 and VCAM-1 in lipopolysaccharide (LPS)-stimulated human endothelial cells'.[4]

OK, so I have no idea what that means any more than you do – except that they're good for you. It seems that poor old Ötzi the Iceman, 5,300 years ago, knew this too, in a different kind of way. Along with his snacking and medical kit, including birch bracket fungus and birch polypore, which he may have been taking to treat his Lyme disease, Ötzi had at least a handful of dried sloes, which he must have picked the previous autumn and preserved, since he died in late spring. Experience may have taught him that they were powerfully anti-inflammatory, for one thing – prehistoric aspirin, as it were.

With sloes, as with so many foods, you are very forcibly reminded of the famous saying misattributed to Hippocrates: Let food be your medicine, let medicine be your food.

If you find even dried sloes a little too astringent for you (though they're mellower than fresh sloes, which can make grown men cry), cooking them up into a sloe and crab apple jelly is an excellent way of eating them.

Then there's the humble blackberry. They're so ubiquitous, the bramble can be a nuisance and that odd sweet/bland/earthy taste isn't always, to be honest, the best of wild fruit flavours. But their nutritional properties just can't be ignored, nor can their ridiculous abundance in the autumn. The wild blackberry is really one of the wonders of English nature.

A single bramble bush can produce nearly 5 kilograms of fruit a year – for free, organic, and arguably with a higher antioxidant content than blueberries. Here are the figures for the powerful antioxidant anthocyanins, one of the main reasons dark blue and purple fruit are good for us, reducing everything from cancer risks to cognitive decline. One study found that farmed blueberries contained up to 26 milligrams per 100 grams, but blackberries can contain a whopping 139 milligrams – although, to be fair, much depends on how the fruit being tested has been picked, stored and so on. On the subject of blueberries, tests always show that wild blueberries have consistently much higher levels of nutrients than the farmed variety. I suspect this is true of absolutely every food you care to mention.

It's good to know that anthocyanins are also very resistant to damage by heat, so stewing up your blackberries or blackcurrants along with apples in the traditional country manner, perhaps adding in a few sloes and elderberries as well, does very little harm and actually *increases* their metal-chelating properties, that is to say their ability to cleanse our systems of heavy metals like lead and mercury.

But if you prefer to expend a great deal of effort digging out those pesky brambles to keep your garden nice and tidy, before driving down to the supermarket to buy a plastic tray of multiply-sprayed yet curiously tasteless blueberries, and at ever rising prices, that's up to you.

Another fruit that cannot be ignored is the aforementioned elderberry. Even though elder wasn't present at the Underhill hedge, it's extremely common throughout the country and hard to mistake for much else – but, as ever, take an expert friend or a good field guide to be sure.

Raw, they're somewhat toxic, though I've eaten one or two and never felt anything. But they're best cooked, and absurdly healthy. For centuries – probably since Ötzi's day, and a lot longer – they've been especially used to keep colds, bugs and viruses at bay in winter. The seventeenth-century diarist and botanist John Evelyn called them 'a catholicon against all infirmities whatever', which is impressive; and the herbalist John Gerard recommends the dried seeds, taken in the morning with a glass of wine, for plus-sized and big-boned persons, or as he puts it rather more bluntly, 'such as are too fat, and would faine be leaner'.

The cold and flu pick-me-up medicine in a bottle called Sambucol contains elder extract, hence the name – elder is *Sambuca nigra* – but why not just make your own? It's a deep purple panacea not to be missed.

Boil up about a pint of water, a third as much honey and some twenty-five elderberry heads. Stir well, cool off and bottle in a sterilised bottle or two, and drink through the winter months to stay absolutely tip top and tickety boo.

Rose petals from dog rose are best eaten raw in salads, but you can also make jams and jellies. And then in autumn you can make the celebrated rosehip syrup, from *any* rosehips, including any variety of garden rose. The key thing to understand is that the seeds inside each hip are covered in small hairs that can irritate the stomach lining, so the process is always the same: bring to the boil, simmer for fifteen minutes, and then mash and strain the liquid through a sieve, and again through fine muslin to make sure you've got all the hairs out. You will be left with a rich liquor, which you heat up again with some sugar (to taste) to help preserve it, and some people add lemon juice as well at this stage for flavour. Bottle and keep it through the winter for its extremely high vitamin C content, which survives this small amount of cooking very well. If you take this daily throughout the winter along with your elderberry cordial, you will have some invaluable protection against those winter bugs and viruses.

As mentioned above, the nutritional values in wild foods are sometimes hard to believe. Here is a chart for comparing the vitamin C in rosehips with that of some other fruits:

Vitamin C per 100 milligrams
Oranges: 53 milligrams
Strawberries: 58 milligrams
Papayas: 63 milligrams
Kiwi fruit: 120 milligrams
Guavas:184 milligrams
Rosehips: 1150–2500 milligrams

No, that's not a typo. Rosehips contain as much as *fifty times* more vitamin C than oranges. And they grow for free. No wonder that generations of country folk have gone out to gather them by the basketful to make rosehip syrup.

You can also make ingenious edible fruit leathers with any hedgerow fruit – a very ancient technique. Stew your fruit, sieve and save the puree, and discard the pulp. Then, melt about a quarter as much sugar into the puree, pour this out into a baking tray lined with baking paper and bake for hours and hours on a very low heat. To be energy efficient you can do this on a trivet on top of a wood-burning stove. Eventually, you will have something resembling, well, fruit leather.

If you want to be *really* palaeolithic, you can add some finely chopped dried meat as well, and some rendered animal fat like suet, to create a complete and incredibly sustaining food – pemmican, as it was called by several Indigenous tribes in North America – that will last for months. There are numerous recipes for this online. I've never made it myself, but a friend who has says that the mix of salty meat, rich fat and sweet fruit makes for a fantastic combination.

Incidentally, many people say our native crab apples are just too tart and sour to be edible, but they're making the mistake of picking them. If you wait for windfalls that have lain on the ground until November or later, you will find they have mellowed considerably and stew up well, especially mixed half and half with regular apples.

And then, of course, there is the vexed subject of hazelnuts.

In the past, people ate a *lot* of hazelnuts. We have found large earthenware vessels and sandy ground pits from Neolithic sites showing that our forebears stored them by the hundredweight; hardly surprisingly, as they make a perfect high-protein, high-fat food to see you through the winter, and come in their own long-life packaging as well.

More recently, in 1778, *Jackson's Oxford Journal* gives an excellent idea of how hedgerows were once viewed as a rich resource for all: 'The present is the greatest Nut Year that has been known for a long Time past; the Hazels are every where bending beneath the Weight of their Produce, and as it were their friendly curves, inviting the Hand of the Rustic to pluck their Clusters.'[5]

They are also frequently mentioned in old country cookery books – but today you'll be lucky to find any by the time they're ripe. Every single one will have been snaffled by the squirrels. We can only assume that in the past, people spent a lot more time hunting, eating and trapping squirrels, partly because they're good eating in themselves, partly because they're such ferocious competitors for the hazelnut harvest. I have spoken to a full-time nut farmer down in Kent about what he does to protect his crop, picturing his trees growing, perhaps, in gigantic chicken-wire hangars or something. But no: he simply shoots every single squirrel in the vicinity – and shoots any other new-comers as well. There is no other way.

For vegans, vegetarians, non-gun-owners and sciurophiles generally, you may have to turn to acorns instead. There's a lot more processing: you will need to grind them down to small pieces and then soak them in cold water for a few days, with regular changes, to leach out the tannins – but you can then grind them into flour and use like any other. They're incredibly abundant, it's easy to gather them by the bagload and, once processed, they are also extremely nutritious.

There's one other nut that is often mentioned: the beech nut. Now, these are also very abundant in good years, but do you realise what size they are? You have to break open the prickly casing and then extract the inner nut, which is often about the size and shape of your typical big toenail clipping. You are then supposed to press this for the oil. Well, I did try one year, and after several hours of collecting, peeling, gently heating the nuts in the oil press (cold pressing is almost impossible) and then squeezing, I ended up with about a tenth of a teaspoon. Unless you have a professional-sized millstone to hand, I really don't recommend it.

Another overlooked tree nut is the ash key, also perfectly edible – but perhaps only one for after the zombie apocalypse because they're not necessarily delicious. Raw, warns John Lewis-Stempel in *The Wild Life*, they taste of wormwood, but he carefully followed a recipe in the Second World War austerity classic, *The Wild Foods of Britain* by Jason Hill, and pickled his ash keys in a rather piquant-sounding liquor of vinegar, cider and salt, boiling and sealing them for at least three weeks, and then sampling with some excitement.

They still tasted of wormwood.

Along with nuts, at this time of year there are also seeds. These may seem so small as to be not worth the effort – even worse than beech nuts. But it's not that simple, since many seeds don't need much processing. Nettle seeds will be hugely abundant where large patches grow, for instance, and you can collect bagfuls without much effort. You should then leave them exposed for a while to let the bugs and spiders crawl out to safety, and finally give the bunch a good shake. Anything left over after that should be consumed as useful insect protein.

You now have a big bag of nettle seed, which can be separated from any fragments of leaf and stalk through a sieve and then lightly toasted in a frying pan. Delicious. Like its trendy close relative, hemp seed, it's full of protein and healthy oils – linoleic, linolenic, palmitic, oleic, stearic, which we make into omega 3 and omega 6.

As with most foods (see broccoli, mentioned previously), too much may have unusual consequences. Nettle seeds in quantity can be an incredible stimulant, causing sleeplessness even more acutely than a dozen espressos or the finest quality amphetamines.

Overall, there are a few golden rules with wild food. Always know exactly what you're picking and eating. Never any one thing in too much quantity, and be imaginative with your cooking. I've never yet got around to making acorn flour, but, when I do, I will make it into some pastry with plenty of butter and lard and a splash of warm water, and probably fill it with a mix of eggs, Cheddar cheese, spinach and walnuts.

Booze

There are some fine alcoholic drinks on offer at the Hedgerow Bar, with a little human intervention. In his indispensable *Booze for Free*, which I cannot recommend too highly to the hedgerow forager and keen recreational alcoholic, Andy Hamilton offers us the recipe for his Boozy Dandelion and Burdock, the two common hedgerow plants usually being made into a pleasant cordial, but here given a little kick. In fact, as the author admits, "Boozy" is something of an understatement, actually. When I made this recipe it fermented to around 9%.'

Incidentally, dandelion and burdock cordial is said to have been invented by the greatest Catholic philosopher and theologian of all, St Thomas Aquinas, who apparently came up with it while out walking and feeling thirsty. You can find this story *ad nauseam* all over the internet, a very good example of how factoids spread in an age of extreme gullibility. But it's rubbish, I'm afraid: a joke concocted by some bored Cambridge students. The Angelic Doctor had his mind on higher things.

But in case you do overindulge on the alcoholic form of dandelion and burdock, and feel a little untoward the next day, Andy Hamilton also offers a wonderful Windy Tea, using nettles along with mint, in case of post-potatory flatus from home-made ale; and Hair of the Dog Tea, involving nettles and meadowsweet among other things. I would also add dandelion and ginger myself.

Make in a teapot with hot water from the kettle. 'Allow to infuse for five minutes. While you're waiting, hold head and say, "Never, ever again."'[6]

You might get less of a hangover in the first place, though, from home-made rosehip vodka. A Scottish farmer's wife has explained on her blog how to make a delicious-sounding version, but really these proportions are much the same with any kind of hedgerow fruit liqueur.[7] And in my experience, the less closely you follow any recipe the better. Just bung it all in.

The rough idea is to fill a wide-mouthed (so you can get the delicious boozy fruit out again) bottle or jar at least a quarter with hedgerow fruit. This can include rosehips, haws, sloes, bullace, crab apples, elderberries, blackberries, blackcurrants and literally any other fruit that you know to be edible. Not all of these are in season at the same time, of course, but that's easy: just top up the bottle with the later ripenings when they're ready. Fill the bottle about another quarter with plain white sugar, more or less depending on how sweet you prefer. Then, fill to near the brim with either gin, for some extra flavour, or vodka, which tastes of roughly nothing at all in my opinion. I have tried brandy, but it tends to make everything taste slightly of cough medicine. Gin is much my favourite. Leave this lot for about three months in a dark cupboard – direct sunlight will cause the taste to deteriorate – giving the bottle an occasional shake, and then strain off the liquid. Remember to do a second strain through muslin if you included rosehips. Store for a year or more. I once made and then forgot about some crab apple vodka, which I drank ten years later. It was superb.

Oh, and don't forget to eat the wrinkly little marinated fruits as well. If you can be bothered, you can mash them all up in a bowl and remove all the stones, then mix the mash with cream. Hard work and fiddly, but it tastes fantastic.

The only other, very ancient alcoholic health drink that must haunt and tantalise the imagination, is the immortal green Chartreuse liqueur, made by French Carthusian monks for the past four hundred years. Alas, we can never replicate its incredible flavours deriving from a mix of 130 or more wild alpine herbs, steeped in some unidentified alcoholic liquor, because it remains a strict secret to this day, the recipe known only to two monks at any one time. But if you tried steeping, say, yarrow for the aniseed, meadowsweet for sweetness, maybe galingale in vodka, you might get something vaguely like it. Or you could just buy a bottle. As Saki's Reginald observed: 'People may say what they like about the decay of Christianity; the religious system that produced green Chartreuse can never really die.'

Hedgerow Foraging in the Past

But what do we know of actual wild and hedgerow foraging in the old days? Did people really eat all these things – and what good did it do them?

They certainly did. When incomes were low, shop food was expensive and free foraged food was abundant, why wouldn't they?

From Alfred Williams' fascinating *Life in a Wiltshire Village:*

> The poor people of the down-side still eat many herbs which would not appear palatable to dwellers in the town, but which, as a matter of fact, are both agreeable to the taste, and highly wholesome too. In this list are stinging-nettles – also used for tea – hop-tops, charlock, boiled lettuce, dandelion leaves and roots, sorrel, wild landcress, and succory.[8]

The children were especially keen on sweet hedgerow freebies on the way to school – in the days when it was quite common for country children to walk four miles to school cross country, from the age of seven and up, as Alison Uttley so memorably describes in her classic *The Country Child.* Instead of sweets and confectionery that nowadays might occupy an entire aisle of a supermarket, country children would raid the hedgerows like a troupe of voracious sparrows.

In springtime, they were fond of the plump, nutty buds and young leaves of hawthorn. Later in spring they would eat primrose and cowslip petals and the tangy, lemony sorrel leaves. It wasn't all confident and knowledgeable foraging, though, and the children would sometimes dig up and try to eat a crowfoot bulb, 'which is remarkable hot and pungent'. In summer, there was the fruit of the maple-tree, known as 'hatchets and bill-hooks', from the distinctive shape of their 'wings'; and then crab apples and wild apples, blackberries, beech and hazel nuts, sloes and hawthorn fruit. The children sometimes even ate yew berries.

They're not poisonous, in case you're worrying, but the stone inside them is deadly. Eating them seems a little risky, but then country children were evidently pretty tough in those days. Alfred Williams records that he himself once ate some toxic berries of lords-and-ladies or cuckoo pint, which 'severely burned my throat ... I wept all the way home.'

A lot of wildflowers we are warned against may be theoretically toxic, but far less concentrated than the synthetic pyrethroids, like lambda-cyhalothrin, now sprayed on those crops, 'peas and beans, turnips and swedes', which Williams' country children used to forage so eagerly. I'm proud to say, though, that I do know a lady in arable country who still goes out 'gleaning' in the vegetable fields after the harvest.

Beyond foodstuffs, the hedgerows and their produce supplied many other things for the rural poor. The bees visited flowers and made honey and beeswax, from which candles were made – but they were very expensive. The poor burned tallow candles, made from the melted or rendered fat off any roasting joint, beef or mutton generally.

But the best light and the commonest for centuries was actually the rush-light, made from rushes gathered alongside streams and ditches, and on our land especially abundant alongside our boggiest and dampest hedgerow. Unlike a tallow candle, the rush-light produced very little if any smoke and gave such a good light that a single one could illuminate an entire dinner party and they were still used into the Second World War.[9]

In his early classic of rural self-sufficiency, *Cottage Economy* (1821), William Cobbett wrote about rush-lights in his customarily vigorous style. And it's important to remember that he wrote during a time of deep agricultural depression and rural poverty, driven by rage at the miserable state of the labouring poor in the villages, while the 'tax-eaters' grew fat in London. He was particularly indignant that poverty had driven many cottage labourers to eat more potatoes and less meat.

'I was bred and brought up by rush-light,' he says, 'and I do not find that I see less clearly than other people … My grandmother, who lived to be pretty near ninety, never, I believe, burnt a candle in her house in her life.' Candles were taxed as well by the London regime – another source of Cobbett's ire.

Cobbett describes how you peel the fattest, best part of a rush of its green skin, leaving only 'a little strip … all the way up, which, observe, is necessary to hold the pith together all the way along'.

Melt your grease or fat, soak the rush in it, then fold it up a curved bit of bark from a young tree, and there it hangs. As Cobbett explains:

> Now these rushes give a better light than a common small dip-candle. You may do any sort of work by this light; and, if reading be your taste, you may read the foul libels, the lies and abuse, which are circulated gratis about me, as well by rush-light as you can by the light of taxed candles; and at any rate, you would have one less evil; for to be deceived and to pay tax for the deception are a little too much for even modern loyalty openly to demand.

Hedgerows also provide free meat, of course, given all the wild animals that use them for board and lodging, from rabbits and squirrels to pheasants. Anyone with a gun or a humane trap and the requisite skill can source high-quality wild game. Hunting with a bow and arrow is now illegal in this country, pathetically: the only country in the world with such a law. But you are allowed, indeed encouraged, to go to the supermarket and buy colourless factory-farmed chicken for your dinner.

But do you realise how much game meat you actually have to eat to get your calories?

A whole squirrel provides about 300 calories. Working all day outdoors, pursuing your off-grid rural idyll, you'll need far more than the feeble 2,000 calories a day currently recommended to a nation of desk-bound, telly-watching citizens. You'll need at least 2,500 – or *eight squirrels a day.*

The truth is, the last time hunter-gatherers lived in Britain, they would have hunted far larger and fatter animals, everything from beavers to horses to aurochs. In tamed, domesticated, synthetic modern life, wild foraging can only ever be an add-on. But many of us find it a consoling one.

Hedgerows for a Premodern Diet

There is plenty of evidence for just how strong and healthy people once were on their pre-intensive agriculture, often wild-food diet.

Certainly, it's safe to say that rural people in premodern times were unbelievably fit and strong, genetic survivors in every way – because if they weren't, they would never have made it past childhood. Even around 1900 in the UK, infant mortality was not far off 50 per cent.

The premodern diet for the poor in the countryside may have been plain but it was highly nutritious, and of course free from chemicals and plastics. On top of that, the amount of exercise they took – *had* to take – was colossal. Anyone today who has spent just three or four hours hedge-laying by hand, or digging out a ditch in winter, as I have, can testify that it is *way* harder than an hour in the gym or an hour swimming. There's just no comparison in how tired you feel, how much you need to eat in the evening or how well you sleep: just as well as you did when you were six.

And if we go back to hunter-gatherers, whose diet was or is entirely wild food, they have famously low to non-existent levels of heart disease, arthritis, diabetes and many other diseases of modernity, while depression also appears to be unknown. Of course, this is an effect of an entire lifestyle – partly wild food, but also a life lived outdoors, sunshine and physical exertion – of which the modern lifestyle is the virtual opposite.

The final conclusions to be drawn from this chapter are very clear. We should be replanting, cutting, laying and generally looking after our hedges as a precious food resource, just as much as the rest of the countryside – but are currently failing badly to do. We should also be permitted proper access to our own countryside – the English people are currently excluded by law from an appalling 92 per cent of their own land. This is a whole subject in itself, beyond the scope of this book, but it is closely related to collecting wild food, as well as a much wider appreciation of nature. A Swedish friend has explained to me that growing up from infancy with an automatic 'right to roam' anywhere in their country without question, while also being instilled with a strong sense of their own responsibilities to that country, gives Swedes a much deeper knowledge of and respect for their land than the English can have. To enjoy the full benefits of wild food, first we have to have access to the wild. But until that time, we will have to do what we can with the 8 per cent of England where we are permitted to tread, to go foraging, cook wild garlic, nettles and blackberries, and pursue as active a lifestyle as possible in our sedentary civilisation.

CHAPTER SIX

Hedges as Carbon Banks

edgerows, like woodlands, are naturally moderating presences in a landscape, having huge potential to mitigate the worst effects of climate chaos. And one of the things they can do supremely well is soak up carbon dioxide, as well as lessen some of the consequences of heavier rainfall and flooding. Hedges and their good management, indeed, could be *far* more important than planting new woodlands, the strategy currently favoured by government (possibly because it is more eye-catching).

Jonathan explains:

'The Government has a target to plant 30,000 hectares of trees by 2025. This target contributes to the UK COP26 goals. But many of these trees are planted in the wrong place and aftercare is either negligent or non-existent. An example of this can be found in a failed plantation in the Lake District National Park, where only some 15 per cent are alive and many of the tree guards are disintegrating. The River Esk runs below this hill and in time micro-plastics will enter the water course. Wrong trees, wrong method, wrong place.

'England has 400,000 kilometres of hedges. If managed well (conservation-style or similar), our hedge network could achieve way more than this misguided planting programme. Our national hedge network is a resource not being utilised – in fact it is being ignored. Yet with intelligent management there would be no need to take existing land out of use, no need to import trees (with attendant risk of imported pathogens, e.g. ash dieback), no need to use glyphosate, a hugely damaging herbicide, and no need to water saplings for the two to three summers after they have been planted.'

Very rough calculations suggest that allowing an established but heavily trimmed hedgerow or 'stump-row' to regrow to 'conservation size and density' could sequestrate about three tonnes of carbon dioxide per hectare per year. The 400,000 kilometres of hedgerow in England equal some 150,000 hectares in area and so might store an impressive 450,000 tonnes of carbon dioxide per year. But the calculation should also take into account things like the capacity of soil adjacent to a hedgerow to store much more carbon dioxide than otherwise as well.

A 2021 study found that hedgerows can increase the carbon storage capacity of neighbouring cropland by 30 per cent or more, and that 'The positive effects of hedgerows can only be fully achieved if they are adequately managed and present in their traditional form. This includes a dense woody structure with shrubs.'[1]

And again, it is vital to realise a very simple fact: our hedgerows are already there; it won't use up or occupy any more land to change them from flailed or neglected stump-rows back to luxuriant linear thickets. As an already existing natural feature that could be a powerful weapon for allaying climate chaos, hedgerows are currently severely underestimated.

Climate-Calming Hedging

First, a note on terms used. 'Climate Change' is now as old hat as 'Global Warming'. What we face is Climate Chaos.

Permanently unpredictable, a nightmare for farmers, growers, gardeners and the entirety of our wildlife, which will no longer experience the weather patterns and temperatures to which it has adapted over tens of thousands of years.

A very simple example: one study found that between 2010 and 2017, the global average wind speed appears to have increased from 7 to 7.4 mph – a rise of about 17 per cent in less than a decade.[2] This may not sound much, but since it's worldwide it indicates a huge increase in the amount of energy there is in our weather systems – not a good thing – and if you're the size of a butterfly, even the smallest increase in wind can make a big difference.

Hedgerows, however, lower wind speed, thus protecting the bare topsoil of much modern farmland. A study of soil erosion between Crewkerne and Yeovil, by the Soil Survey of England and Wales, found average losses of 4.2 tonnes of soil from every winter cereal hectare, though one field suffered an appalling 21 tonnes of soil per hectare.[3] Increased wind speeds will only exacerbate this dustbowl effect in the future – but hedgerows can help.

They also soak up excess rain and reduce flooding. Another effect of climate change is far higher but more abrupt and concentrated rainfall, so we should also note that, as the excellent campaign group Hedgelink have calculated, a 50-metre stretch of hedgerow at the bottom of a field 'can store between 150 and 375 cubic metres of water during rainy periods for slow release down slope'. Furthermore, 'This effect is greatest in soils rich in clay or organic matter. [And] because of their deep roots, hedgerows remove water faster from the soil than crops during periods of excessive rainfall.'[4]

This applies to a dense traditional hedgerow, though. The modern stump-row will often be adjacent to a soil surface baked hard in summer and compacted by gigantic modern tractors, like the Case IH 620, which weighs a staggering 24.4 tonnes, popular with East Anglian farmers. Soil surfaces compressed by this behemoth will only produce more run-off in heavy rain.

John Dover sketches out some further, overlooked eco-services of hedges: soil formation, the production of oxygen through photosynthesis, nutrient cycling, water cycling, air-quality regulation from the removal of particulates and nitrogen dioxide, the prevention of agrochemical drift, and many more.[5]

Two other points: regular hedge-laying and trimming produce a vast amount of brushwood and cuttings, which could easily be converted to woodchip for biofuel at extremely low cost. Indeed, for centuries, as we mentioned earlier in the book, the hedgerow was a source of fuel for country people as well as food: the 'bundle of faggots on the hearth' was precisely a bundle of hedgerow cuttings (or coppice cuttings) taken in winter, often tied with dried nettle rope or willow withies. These also dry much more quickly than logs and therefore could be burned sooner, and were infinitely renewable.

Hedgerow Corridors

So: hedgerows and climate chaos – how can they help? One of the very simplest ways lies in that rather dry word 'connectivity', or the rather more evocative phrase 'wildlife corridors'.

It seems an odd idea. Why can't wildlife just cross open countryside? But this is to think as a human being, a large mammal who has long since exterminated any of his potential predators from the landscape. Things appear very different for smaller creatures, which rely on hedgerows for concealment from predators as they move from one side of a field to another, say. Hedges may also connect two small woodlands, a woodland to a lake, and so on. Dormice, in fact, hardly ever come down to ground level if they can help it. Offering protected highways and trunkroads across our landscape, as it were, hedgerows make it much easier for wildlife to move around safely.

There is also a prevalent idea that certain native trees will be more resistant to abrupt climate alterations than others. Jonathan Thomson reports from observation that certain trees and shrubs flourished during 2022 at Underhill, despite the prolonged drought and record temperatures – in particular gorse, a very valuable shrub for insects and pollinators. Other species that are less susceptible to the effects of climate change would include oak, aspen, wild cherry and sweet chestnut. But I must admit, swapping our native blackthorns and hazels, say, for sweet chestnut, seems to me an admission of defeat, with so much of our native wildlife dependent on the former, not the latter.

Dr Rob Wolton is a great cheerleader for hedgerows in the fight against climate chaos and its multiple associated, sometimes unforeseeable problems. In his 2019 paper 'Hedges – How They Can Help Us Respond to Climate and Extinction Crises?', he reminds us that there 107 priority species for conservation associated with hedgerows and their 'value for nature conservation is out of all proportion to the area of land they occupy'.

Meanwhile, the CPRE, The Countryside Charity reminds us that the number of ecosystem services rendered by our hedgerows is twenty-one; that of woodland is nineteen, and flowering grassland meadow sixteen.

Wolton also says that in Brittany, 13 per cent of the carbon in the landscape is contained in hedges, so you can reckon it's *at least* this much in the well-hedged UK. 'The Climate Change Committee has called for a 40% increase in hedges to meet Net Zero by 2050,' he points out, 'but this doesn't specify management which is crucial.' Dense hedges are what we should picture.

Meanwhile a newly planted hedge, though less help for wildlife directly, may store 600 to 800 kilograms of carbon dioxide per year per kilometre, for up to 20 years; another study estimates an 'annual 3.3 tonnes of CO_2 per hectare for the first 50 years' which comes to about the same. The very maximum amount of carbon stored per hectare (not kilometre, note) is a staggering 350 tonnes though 'the average is far less, around about 4 tonnes per hectare'.[6]

For comparison, the average family car in the UK emits 1.6 tonnes of CO_2 per year. Although since there are around 120,000 *more* of them on our roads each year, that means 192,000 more tonnes of emissions. Hedgerows can't do everything.

One of the most interesting experiments by a farmer has been that by Ross Dickinson in Dorset. It suggests a whole new way in which hedgerows might help us reduce carbon dioxide emissions and fight climate chaos, by providing an endless supply of *almost* carbon-neutral firewood, and, crucially, in a way that is profitable to the farmer.[7]

Farmer Dickinson changed from flailing his hedgerows to a longer cycle of coppicing with a tree-coppice machine. That is, simply cutting certain selected hedges back to ground level. It looks brutal, as if the hedge has vanished altogether for some months, but then it sprouts back from the old-established roots with amazing vigour. Indeed 'young shoots on old roots' is an ancient description of coppice woodland. The result is that the hedges around the farm are all at different stages of growth, which is great for biodiversity, as well in much healthier condition. Rather than simply growing out and growing old, they are constantly rejuvenated, as all coppice woods are.

Dickinson did the following sums: 220 metres of hedge, mostly sycamore and ash, were coppiced back to the ground, produced 9 tonnes of saleable logs and six tonnes of 'ugly sticks' for domestic burning, as well as 260 bags of kindling. Dickinson estimated that even after labour and fuel costs, he made £1,530 profit from the sale of firewood from this length of hedge alone, which can be repeated every fifteen years. Multiply this by the total length of hedges on his farm, and it adds up to a very tidy sum. The total firewood produced from this one 220 metres of hedge equalled 88 megawatt hours of heat energy, or 8,500 litres of heating oil, *and* was very close to carbon neutral.

If Dickinson had also claimed subsidies (which he didn't), he'd have been better off still.

Now admittedly, cutting a hedge right down to the ground is a long way from the conservation-laid hedge at Underhill, which in our dreams would become the new normal across the countryside. But failing this, where costs seem simply too high or the skills are lacking, mechanical hedgerow coppicing on a regular cycle like this seems to promise huge benefits too.

Bill McGuire, professor of geophysical and climate hazards at University College London, points out that when *Homo sapiens* first appeared on the scene, they shared the planet with an estimated six trillion trees. Today, thanks entirely to the activities of this busy little simian, the number is down to three trillion. Since trees in these numbers have a huge moderating effect on planetary climate, the consequence of getting rid of half of them are, to say the least, substantial.

But a creature named *Homo sapiens* would, of course, have the wisdom to call a halt at this point, yes? Especially when threatened with climate change so severe and sudden that it endangers not just our current civilisation but possibly our existence?

But no doubt you know the answer already. Along with all the other eager contributions we continue to make to ever more dramatic climate chaos, with all the enthusiasm of a simple-minded squirrel throwing itself into a cement mixer, we continue to destroy a further sixteen billion trees *every year*. The tropical rainforests alone 'are being shorn at a rate of 476 trees every second'.[8]

Will planting hedgerows, or relaying and regrowing the bleak, denuded hedgerows that we already have in this country, cancel out the tropical decimation? Definitely not. But might they help a little, along with other things like biodiversity, soil richness and so on? Yes, definitely.

So let the planting, laying and coppicing begin!

CHAPTER SEVEN

Pleachers, Bodgers, Brummocks and Flailers

Hedge-Cutting, Ancient and Modern

The art of clipping and laying hedges is carried to great perfection in England, and great encouragement is given by some landlords to encourage emulative skill in this department of labour. Sir Gilbert Heathcote, Bart., for example, has given prizes, from 5l. to 2l. each to the best hedge cutters and ditchers in Rutlandshire.

Martin Doyle, *A Cyclopædia of Practical Husbandry and Rural Affairs in General*

T he rarest, most visually pleasing and wildlife-rich hedge is one that has been wildlife or conservation laid – with the traditional, tightly bound Midland bullock hedge coming a close second, after a few years of regrowth. But be aware, the reality of actual hedges and hedge-laying may not always fit these neat names and categories.

How is it done?

In his own invaluable guide to rewilding small areas of land, *How to Rewild*, Jonathan gives this concise summary of how to go about conservation hedge-laying – although these days, he prefers to call it 'rough hedge-laying' to reflect the intentionally untidy style.

1. Begin at the end with the highest elevation.
2. Before starting, you may need to clear a space to lay into. Given the need to lay uphill, this area will be the highest within the run you are going to lay.
3. Lay the hedge to lie uphill – sap works from the roots up the plant. If you lay dead flat or downhill, you are likely to kill the plant.
4. Cut up to 75 per cent of the way through the trunk at about 18 to 24 inches from the ground. Then gently lay the plant over, never going beyond about 10 degrees of incline.
5. Move on to the next plant and repeat: cutting and laying.
6. If you need to remove the lateral branches to gain easier access, then do this.

Good clear instructions – and you will learn more again by experience. I know people who have done a 'weekend course' in traditional and complex styles of hedge-laying, which is certainly a start, but really it takes weeks, even months, before you develop real skill in this. With the quick and simple method of rough hedge-laying that we are advocating, however, you can become proficient very rapidly, and Jonathan reckons he can do as much as 30 yards in just one morning.

You will also learn that every species of tree behaves slightly differently, recent wet or dry weather can make a difference, working in November is quite different to working in February, and then there's the wind direction to consider and the slope of the land, as laid trees should be leaning slightly uphill. And on top of all that, you need to make sure you don't cut your thumb off with your brummock.

('What do you call this tool?' I asked old Dennis the Shropshire sheep farmer. 'I'd call it a billhook.'

'I'd call it a brummock,' he said. 'Though it comes to the same thing, I suppose.')

There are a bewildering variety of regional styles in traditional hedge-laying:

Lancashire and Westmorland;

Midland bullock;

Leicestershire bullfinch;

Devonshire;

Welsh border;

and so on.

Much depended on what use they were intended for, and you can pretty much picture how they would have to look if you consider the typical farming in each area. Lancashire and Westmorland? This must mean sheep country, surely, and a hedge style dense enough to contain these woolly wanderers, who will always want to get into any field other than the one they're already in. The same might be assumed of the Welsh border hedge.

The Midland bullock, meanwhile, demanding a lot of skill and practice in creation and laying, is a tightly bound wonder, immensely strong, enough to stop a bullock – which was precisely the idea, the rich lowlands of Leicestershire, Northamptonshire and Warwickshire being traditional beef-raising country.

The uplands of Yorkshire have always been more sheep country again, so where it wasn't a drystone wall doing the job, the Yorkshire hedge was low but very dense and thorny to stop any of the woolly stock squeezing through and going off for a wander – as sheep so very much like to do. You will find the same sort of hedge in other sheep country, such as the uplands of Devon.

On top of that, there is a fantastic vocabulary to go with it. As well as billhooks and brummocks, there are pleachers, plashers, beavers (also known as frith in Surrey, if that helps), edders and winders, burdens and cags and shards, burrs and pares, stubs and ditching...

Marvellous and slightly daunting as all this sounds, be not afraid. Beneath all the tasty dialect terms and regional styles, the essence of the ancient hedge-laying art remains the same – because trees respond to cutting in roughly the same way.

Here is how you go about a traditional style of hedge-laying: first, you will clear the ground so you can actually get to your hedge. You need access to the trees and stems you are going to cut as near to the ground as possible, so you may have to clear away quite a lot of the brash or brushwood. In conservation and rewilding you may pile this up to form a woodpile for wildlife or distribute it under the hedgerow to rot down. Or if you're in Powys, you may shove it back up against the cut stems at the end of the process to protect from rabbits and others, in which case you will refer to it a *byrdn*.

Now you can see clearly which of the stems to remove altogether and which to cut and lay. The ideal size to lay will be about the circumference of a man's wrist, but there is a lot of leeway and one of the acquired skills of the experienced hedge-layer is knowing how old a tree can still be laid.

Take your billhook or brummock and make a diagonal downward slice into your stem, variously known as a pleacher or stolling, and bend it carefully over, away from the stub. It should be laid at about a 30-degree angle, and the remaining shard of wood should be trimmed with a handsaw. Then onto the next one down – and keep going.

At some point, you will also incorporate stakes from wood cut from the hedge itself, the point made with some sharp axe strokes and hammered in alongside your hedge on both sides. Finally, you will add the aesthetically pleasing plaits, or binders, edders or winders; of hazel ideally, being the most flexible, or maybe willow, knocking them down around the stakes along the top to hold everything in place and stop your stems springing back up, as they will try to do when the sap rises in March. You may also add some crooks at strategic places; a crook being a long stick with a sharply hooked top, whacked down into the ground with a mallet to hold pleachers in place.

Finally, stand back, admire your handiwork, and have a cup of tea.

Job done.

And it should last at least twenty or thirty years.

But as I say, it can take weeks, months, arguably years, to create a truly lasting laid hedge, and it is a thing of beauty to behold. In the very first summer it will sprout back up and thicken with astonishing vigour, and within three or four years it will be absolutely filled with nesting birds. Once again, a feature of the countryside designed to feed or support human beings turns out to be loved by wildlife as well. Other great examples would include the hay meadow, the village pond and the mill pond, the coppiced woodland, the farmyard barn, the drystone wall and many more.

The Most Wildlife-Friendly Hedge-Laying

The procedure with conservation or rough hedging is far simpler, as we have said. There are no stakes, no bindings and very little brushwood, as you leave all the shoots and lateral growth in place to produce the dense, scruffy mass of a hedge with wide skirts that we are looking for. It will be a lot more ragged and thickety, sprouting out in all sorts of unlikely directions, perhaps with some taller shrubs left along the top here and there, and standing dead wood too.

Just as important is the field-side profile: it shouldn't have straight lines and, if laid by an amateur wildlife-loving hedge-layer and bodger (like Jonathan, Harry and me) rather than a professional, it won't have. It will have a naturally wavy or 'scalloped' edge – not to say downright messy – rather than the hard edge of a machine cut. On a tiny scale, such a hedge is like a coastline, a continual flow of inlets, bays and head-lands, forming invaluable microclimates and wind-shelters in themselves, especially loved by insects like butterflies on warm summer days.

Unlike with traditional laid hedges like the Midland, a strong and very tidy style of thick, clean interwoven stems, with wildlife hedging the side shoots are always left on the pleacher as much as possible, meaning less trimming work required and an ever-thickening hedge spreading out across the land. This is exactly what most wildlife wants: the wider the wilder. Nor should any of the cuttings and dead wood be cleared away and burned, as they are in 'tidier' farmland styles. Instead, they should be piled up alongside or underneath the hedge, repli-cating the natural process of falling twigs and branches. There they will gently rot down and provide endless food for beetles, grubs, fungi and larvae as well as marginal hedgerow plants like garlic mustard, hogweed, burdock and that great lover of rotting wood, the stinging nettle.

Regarding the slightly messy and uneven appearance of a rough wildlife hedge that looks rather like the work of a drunken barber, as spiky and unkempt as Hagrid's hairstyle, Macdonald and Gates (in *Orchard*) brilliantly remind us that all our native tree species evolved to grow alongside both straight-tusked elephants (around until about 35,000 years ago) and beavers – 'two of the greatest tree fellers of them all'.[1]

Then there were the colossal blunderings and bulldozerings of aurochs, the browsing of deer, the gouging by their antlers, the nibbling by hares and voles… not to mention the myriad microbial attacks by bacteria and fungi. In response our tree species have evolved an array of defences from the obvious – thorns – to the toxic.

For the amateur and slightly clumsy hedge-layer, this is all good news. Have no anxieties about slicing into the trunk or pushing it over. You're only doing what the great, lumbering, forest-shredding beasts of primeval Britain have done for millennia with their 'rampant herbivory'.[2] When laying a hedge, you're getting in touch with your inner aurochs. The trees can take it and indeed will bounce back with sprightly energy: precisely the energy that goes into creating the hedge.

A newly shooting hedge is really a vigorous display of our trees' and shrubs' natural and long-evolved response to being savaged by elephants and aurochs. Although sadly hedge-laying can't replicate the effect of herbivores in one respect: 'The enzymes in an animal's saliva trigger even more dramatic bifurcation of branches and the production of powerful chemicals and thorns, as well as "emergency flowering",' say Isabella Tree and Charlie Burrell in *The Book of Wilding*.[3]

I suppose you could try spitting on the cut ends of your hedge trees and shrubs. But your neighbours may think you even madder than before.

Tree and Burrell also tell us that naturally regenerated ancient woodland stores carbon dioxide six times better than agroforestry and forty times better than plantations! Managing and nurturing an ancient and established hedge is surely a good proxy for such regenerated woodland. 'Woody shrubs too, invest heavily in their carbon-storing root system when their leaves and stems are browsed.'[4]

When worrying about 'untidiness' in hedges or elsewhere, we also have a tendency to assume that trees should grow upright, giving that well-known cathedral-like effect to a beech wood, say. But in a truly natural wild woodland, like the Great Smoky Mountains National Park in Oregon, as Oliver Rackham has pointed out, or in the Białowieża Forest in Poland, as Jonathan himself has observed at first hand, many of the trees are leaning sideways at crazy angles, and some are horizontal altogether, yet still growing happily enough. This too reminds us of a laid hedgerow, a 'horizontal woodland' in miniature, with multiple shoots, and actually a far more ancient and natural form of woodland than a uniformly serried upright one.

But is it realistic to hope for more hedge-laying by hand? Even by volunteers? If so, how much? Can this be done on sufficient scale to make a real difference to our environment? Or – most tantalisingly of all – might there be other ways to combine mechanical speed and power with wildlife-friendly hedgerows?

The enormous power of fossil-fuelled machines has changed our current management style of hedgerows completely, as it has so much else. A modern flailing machine will cost a farmer around 88 pence to flail a metre of hedge or £13.20 per metre over 15 years, 'in diesel, machinery costs including wear and tear, depreciation and labour', according to the Organic Research Centre 'Hedgerow Harvesting Machinery Trials' report.

This is, on the face of it, far cheaper than traditional or con-servation hedge-laying by hand – as with any other farmland task you care to mention. You can imagine the cost difference of digging up, removing and burning all the docks from a field by hand – as was done within living memory – compared to simply spraying the lot with a targeted herbicide.

As a result, food is now very, very cheap compared to average incomes – but as many environmental campaigners have pointed out, this will prove short-term and unsustainable, as we steadily exhaust nature's ability to renew itself, and meanwhile food is only cheap *at source*. The costs are borne elsewhere. Because we enjoy cheap food, for instance, we have to pay higher water bills to cover the costs of cleaning up our rivers contaminated by slurry and run-off and removing agricultural chemicals from our drinking water. We also have to pay higher taxes to fund colossal government initiatives to try to reduce climate change – vast tree-planting projects, for example – while farming itself produces an estimated 10 per cent of the UK's carbon dioxide emissions.

Modern, intensive, unsustainable farming is not the 'fault of farmers', by the way: it's the way farming is obliged to be in response to supermarket pressure for cheap food. There is also a system of overproduction that resulted from the Common Agricultural Policy (CAP), which incentivised farmers to produce more food, that is still being addressed today.

Better land management, including hedgerow management, could help on every front. Letting our hedgerows simply grow larger, more varied in shape, size and cutting routine and, where possible, relaying them in a traditional manner, would have enormous benefits.

Meanwhile, as regards the dominant current style of hedgerow cutting, totally reliant on fossil fuels, some farmers at least do evince a growing dissatisfaction, although feelings are mixed. Here are some opinions from the front line as to what is and isn't thought possible, and what the future might hold:

Does anyone else really wonder about the future of annual hedge trimming? Or is it just me? The sheer cost of doing it now, £45–50+?/hr, it does make you wonder.

We used to do some 650 hours a season of it, we're now about just over 400. Lost one farm to rewilding, others have gone elsewhere due to timing and cost. I dunno; I can see some just going with it for this year, but it certainly isn't going to get any cheaper. We've priced up a new trimmer for next season and the cost of the thing is absolutely eye watering.

It's one of those swings and roundabouts things. It costs me about £5,000 per year in contracting charges to have my hedges trimmed, and that's not the whole farm each year, that's roadsides every year and the rest every other year. If I left it to grow I'd save some of that each year. Not all, because if I left the roadsides, I'd soon have the council on my back, so I'm going to have to pay for that regardless. But I could probably cut 7 per cent off. But then, after ten to fifteen years, I'd have massively overgrown hedges, I'd have lost God knows how many acres to hedges spreading into the field, and then be faced with how to manage the out-of-control growth. And that's going to probably be as expensive as the annual flailing bill was.

Fifteen years ago, when I took over my place from my father, the hedges were in a right state, having been largely left for twenty years prior to that, with just occasional side trimming to reduce the worst of the sideways spread. It's taken me over a decade of working each winter to get most of the hedges back under management. That's a lot of hours on the excavator with a tree shear, a lot of chain sawing, a lot of hours on the telehandler pushing up and burning brushwood and generally tidying up to achieve it. Some of it is my time, some contractors (the chain-sawing mainly), all of it my machinery. But I'd have put that annual cost at somewhere in the same ball park as my hedge-flailing bill now.

So, it seems to me that you could take a holiday from hedge cutting, save X thousands of pounds per year, only in ten to fifteen years' time to have to spend the same amount of money per year sorting out the resultant mess. I guess that the savings in the here and now could make a difference if times are really hard. But the reality is you're just skimping on main-tenance which will cost you eventually.

Our hedges are going into a management scheme next year where they will only be cut every two years. Initially I thought, yeah, that makes sense, however since attacking some two-year growth compared to one-year growth, I personally think the structure, shape and health of the hedge is better if done yearly.

Two-year-growth gets leggy and doesn't cut as nice. The hedge is now open and not great for nesting birds... Personally, I think this is actually a bad thing.

As you say, it leads to open growth, which makes poor habitat for small hedgerow birds that are supposed to be struggling the most. It's only pigeons that gain from these ideas.

I've noticed that since I coppiced a lot of the larger stuff in my hedges and instituted a programme of regular flailing once everything had regrown for two to three years (every year for roadside hedges, every other year for internal ones), the amount of new growth is amazing. I also think that regular flailing is better for the hedge than allowing it to grow for too many years and then hacking it back – flailing small branches means the wood shatters and dies as far back as the split and takes far longer to recover, whereas flailing new shoots just cuts them off clean and they regrow very fast from there.

The last comment here is perhaps the most interesting. Where our hedgerows have grown out badly into mere stumps or lines of young trees, then coppicing them completely down to the ground would be a great response, quickly producing masses of thick young growth off the old stools: a terrific wildlife habitat. And it's important to note how many farmers think an annual cut keeps the hedges not just neat and tidy, but tighter and better for small songbirds, while trying to cut a hedge after two or three years can be very hard on the machinery – which is expensive – and cause a lot more splintering and damage to the hedge as well.

As another farmer wisely put it to me, 'Hedge cutting is a bit like clothes: one size doesn't fit all.'

Diversity of hedgerows is key. And this includes diversity of hedgerow-management styles, among which wildlife hedge-laying, as with the Underhill hedge, should play a key role.

It's exciting to realise, though, that a *kind* of rough wildlife hedge-laying can also be done with big machines, proceeding much faster and getting more done, in the style pushed by the Natural Environment Research Council and others. Big machines are still going to compact soil, burn diesel and so on, but you can't have everything and hand-laying all our 400,000 kilometres of hedgerow remains a pipe dream.

In their report on Rejuvenation of Hedgerows, NERC reckoned that cutting one hundred metres of hedgerow with a circular saw cost £166, or £1.66 a metre, about double the cost of flailing, quoted earlier in this chapter.

Traditional hedge-laying, though, costs £1,241 per hundred metres, or £12.41 a metre, some ten times as much, and also takes 33 minutes per metre.

However, if hedge-laying were done by *properly skilled* volunteers, not just enthusiastic bodgers (who will drive farmers away in disgruntled droves), a lot more might be achieved.

NERC defines the process of mechanical wildlife hedge-laying thus:

> Each stem was partially cut with a chainsaw and a mechanical digger used to push the hedge over along its length. No woody volume was removed, and some stems were entirely severed when the hedge was pushed over. The average width was more than twice that of the traditional and conservation laid hedges.

> Three years after rejuvenation, wildlife hedging resulted in slightly less vigorous woody re-growth

than traditional hedge-laying, but a greater density of woody material with smaller gaps in the hedge base. Immediately after rejuvenation there was more dead wood in the hedge (up to 40%) due to the severed stems. Berry provision was as good as from hedges that were not rejuvenated.

They also note that 'reducing the cutting frequency from every year to every three years resulted in 2.1 times more flowers and a 3.4 times greater berry weight over five years.'

The cost is intermediate: £413 per hundred metres, but with suitable subsidies and encouragements, some landowners might well adopt it. I'd question whether it would be stock-proof though – especially for sheep! An additional fence would surely be needed as well. But the wildlife benefits and wider eco-services could be enormous.

The other very eco-friendly option, taken in the longer term, is coppicing an entire hedgerow and letting it grow again, as has been done by Ross Dickinson, described in the previous chapter, resulting in a very substantial crop of firewood. If done in rotation, only some hedges at a time, this could be very effective.

There are, then, some encouraging ways in which we can have a great diversity of hedges and hedge-laying styles, cut and managed by everyone from volunteers with their sharpened brummocks at the ready, to farmers with their tractor-mounted circular saws and flailers. However, many firmly believe that we should move away from flailing altogether. Jonathan says: 'Annual flailing brings little ecological value. It strips out berries, nuts, and nesting and roosting opportunities. It also tends to leave the base of the plants open to wind and rain, which creates an inhospitable habitat. Flailing on a three-year rotation brings reasonable ecological returns as there will be three years' worth of those hedgerow fruits and bird nests. But hedge-laying, by some margin, delivers the greatest gain, and conservation hedge-laying is best of all. The exception to this is the management of roadside hedges, which have to be cut more regularly for road safety reasons.'

The worst option, though, would be grim indeed: a countryside of sad, remnant hedgerows, or stripped of them altogether, and replaced with barbed-wire fences as the 'cheaper' option. But this is a perfect example of hidden costs in action.

A Barbed Option

Fencing wire topped with barbed wire is the farmer's quick and easy fencing of choice. Barbed wire is made from steel, which requires temperatures of around 1500°C to manufacture. Even recycling old barbed wire into new will require temperatures over 1000°C. And where does all this energy come from? Overwhelmingly, fossil fuels. Even if it's electricity from wind turbines, those still produce 11 grams of CO_2 per kilowatt hour, because of the manufacturing process. Wind turbines, solar panels, nuclear power stations, all continue to exacerbate climate chaos: they just do so more slowly than oil and gas.

The trouble is, laying a hedge with chainsaws and diggers also burns fossil fuel. Manufacturing a bicycle burns fossil fuel. Almost everything in modern life burns fossil fuel. But given all the other benefits we have covered in this book, a hedge should still be favoured over a fence.

Quite apart from anything else, the actual work of planting and nurturing, relaying and cutting a hedge is intensely enjoyable, while wrestling with barbed wire and a pair of long-handled wire cutters is no fun at all, as I can testify. And then when your hedge begins to feed you and to home countless species of insects and birds into the bargain, there's no comparison.

Barbed wire isn't just less wildlife friendly than the traditional hedgerow, it can be lethal. I've had to unwrap a still-living barn owl from a barbed-wire fence, its snow-soft wings caught up around it. It reminded me horribly of soldiers in the First World War similarly trapped. 'Have you seen the old battalion, hangin' on the old barbed wire?' as the Tommies used to bitterly sing.

The RSPB man took it away in a box but said sadly that it wouldn't live. Not once its wings were broken like that. He also told me that deer and even foxes can be caught on the barbs, dying slow and agonising deaths of hunger and cold. Quite a contrast, then, to the living world that is the thick, tangled old English hedgerow, bustling with life.

Let's Get Scruffy

But this is a not a book of doom and gloom; it's a call to action. The art of conservation hedge-laying is not in itself difficult to learn, there are up to 400,000 kilometres or more of hedge awaiting, and the results are quite spectacular. Remember that one figure from Harry's dissertation, cited in the introduction: that a conservation-laid hedge can increase its foliage density from 20 per cent to 100 per cent in just four years. The potential this implies is incredible: 400,00 kilometres of hedgerow, quadrupling in foliage density.

In reality, some farmers would be against this, and nowhere near such an amount could be thus laid anyway. There isn't the workforce. But it remains an inspiring and tantalising image.

Our monocultural countryside has been devastated by a catastrophic 'narrative of simplification', as Macdonald tellingly puts it, but dense, tangled hedgerows mess up and complicate these boring, linear, tidy, flattened-out and lifeless zones, as do boggy old ponds, ditches filled with rotten leaves, scattered cow pats drying and crusty in the sun, stands of dead trees and the bleached bones of fallen timber lying in wind-stirred grasslands.[5]

Today there are people who have recently moved to the countryside, I'm sorry to say, who then complain about the horse poo in the lane outside their dream cottage. Sigh. We need more herbivore poo around, not less. Tons more. I have spoken to an old countryman, in his eighties, who was genuinely nostalgic about the number of dungheaps that lay around our village a generation ago or more, and an absolute magnet for insects and birds.

He was pretty scathing about vegetarians. If we all went back to eating more mutton and beef and cheese, instead of chia seeds and avocados, he reckoned there'd be far more wildlife. He remembered when there were still horses doing the rounds in the lanes, carrying the milk from the farm, delivering locally right into the 1960s. The lanes had far less motorised traffic and far more birds: he recalled the excited swoop of martins and swallows as they feasted upon the high-protein clouds of invertebrate life that arose from the dungpiles all along insect alley, between the high, dense West Country hedges.

It must be understood, the ecological collapse is not some time in the future. The ecological collapse is now. We're living through it. The most rapid mass extinction in earth's long history, driven by that ambiguous, tragicomic accident of nature, 'the rise of intelligent life'. Us.

But given that we are reasonably intelligent and reasonably rational, this freakish aberration of *Homo sapiens* might yet learn to turn its intelligence towards nurturing and expanding life on earth, not impoverishing it.

We need to mess things up, to reintroduce creative destruction as if by aurochs, to rewild our country where possible. And we can do that without using up much more land at all, by relaying and rewilding our hedges.

Future Hedges

We are living through a human-made mass extinction, the sixth or possibly the seventh in earth's history. How do we face up to this and keep going, and how on earth can hedgerows help?

Well, hugely complex events like rapid climate change or mass extinctions are, by their very nature, unpredictable. And it's the unpredictability, of the world we live in, of the coming century, and beyond, that can give us our freedom and consolation.

One can even take some sardonic comfort from the number of eco-doom pronouncements that *haven't* kept to their supposed schedule.

In 1989, the UN warned that entire nations could be underwater by 2000: didn't happen.

In 2004, the *Observer* reported that by 2020 Britain could have a Siberian climate: didn't happen.

Also in 2004, Al Gore said that it was very likely the Arctic ice cap would be gone by 2016: didn't happen.

In 2009, Prince Charles warned we only had eight years to save the planet – but we're still here, and he himself has had a promotion.

Things are indeed pretty bad – extinction-level bad – but we do still have a choice. And as was shown by the starting point for this book – Harry's dissertation on the Underhill hedge – a countryside rich in conservation-laid hedgerows alone could

178

play a key role in mitigating climate chaos, and increasing insect numbers several times over as well. And if the insects are flourishing, most other things should do OK too.

Preserving hedgerows, along with ancient woodlands and wetlands, grassland and oceans, is always going to be the right thing to do. It's always right to respect life on earth, no matter what may come – and to be honest, no one knows *exactly* what will come, or when, or what the outcome may be.

'Think global, act local' is now a well-worn eco-phrase that dates from at least the 1970s, but it remains as true as ever. It's a deep joy to be able to respond at the simplest and most physical and practical level, billhook and pruning saw in hand, to complicated global anxieties. Jonathan and I have learned this from direct experience, working on and observing the wildlife flourish and grow on our respective rewilded patches. Even seeing the steady spread and sprawl across my three fields of lovely, dense, fruit-laden bramble clumps and thickets, the bane of tidy gardeners everywhere, has filled me with a powerful sense of nature's indefatigable resilience.

But it's my relaid hedgerows that give me the greatest satisfaction, for the simple reason that of all the various micro-environments on my seven acres, from young woodland to old grassland, from chuckling riverbank to permanent bog, it's the hedgerows that feel most magically alive. Similarly with the Underhill hedge: it's a crucial component in a complex interactive mosaic, and in springtime, with the dawn chorus, you strongly feel that nowhere is quite as vital as the hedgerow.

It remains a drastically underestimated ecological niche, its potential constantly unrealised. Yet there are signs that in our countryside and on our farmland, farmers are beginning to tire of the time and money that they have to spend on annual hedgerow flailing. The danger is that they will just abandon or remove their hedgerows altogether – but there's also an opportunity for our hedgerows to be treated very differently, and once more valued highly, as they used to be – and that includes people, perhaps teams of well-trained volunteers, cutting and relaying them in the traditional manner.

If the day comes when the people of this country are allowed to walk freely across their own landscape, rather than being confined to strict rights of way, and learn again to forage widely from woods and hedgerows, attitudes could change even more, as we realise what a resource this thorny food bank can be.

Meanwhile, the best place to tackle our 'hedgerow-blindness', as we might call it, could be with schoolchildren, and starter projects to rewild any hedges on their school grounds or around their playing fields. I wish we could have done so when I was at school. I can still remember the boring shrubby hedge around our playground: dark glossy leaves with white berries, which looked like they might be edible but tasted disgusting. I now know it was a snowberry, from North America, and the berries are actually poisonous to humans, though I don't remember any of us succumbing. And being non-native to the UK, it was pretty useless for wildlife as well.

How many schools, not to mention universities, public parks, libraries and offices, maybe even allotments, have similarly useless or monocultural hedgerows around them? And how rewarding it would be to tear them down and replace them, or at least broaden them out and mix them up with some lovely native hawthorns or spindleberries, some elegant hazels or pale green sea buckthorns, finally adding in some bramble cuttings and honeysuckle too.

Come spring and summertime, such a hedge would be a nature lesson in itself for schoolchildren, from which they might learn the difference between a red admiral and a peacock butterfly, coming to visit the bramble flowers, or the difference between the song of a robin and a dunnock. And replacing a non-native hedge with a native, or else introducing far greater complexity and variety to some boring monocultural suburban beech hedge, would, unlike many environmental projects, take up no more space, and demand very little time.

For the truly ambitious, such a hedge might in time be cut and laid into the kind of magnificent living edifice that we have seen take shape at Underhill: an example for anyone who wants to begin rewilding on a modest but hugely satisfying scale.

If there *is* room to create a new hedge altogether, then nothing could teach greater or more important lessons than a sunny, ideally south-facing, 'edible hedge'. I created such a thing from scratch only a few years ago around our vegetable patch, comprised of various bushes of blackcurrant and redcurrant, gooseberry and raspberry and a couple of exotic Japanese wineberries: non-native, yes, but adored by blackbirds and thrushes, and by the roe deer who hop in over the gate from the field especially to eat the tasty young shoots in the spring and the fruit in summer.

And yet that was always the idea: our edible hedge, like every rich and flourishing countryside hedgerow, with its due complement of blackberries and sloes, hips and haws, is meant to provide food for all, not just for human beings. And it's been so successful, fruiting so vigorously thanks to plentiful addition of home-made compost, mulch and wood ash in winter, that we still end up with tens of pounds of soft fruit every year, which we freeze and continue to eat right through to the following spring.

Establishing an edible hedge makes for a wonderful learning experience and teaches a key lesson, especially when absorbed in the form of blackcurrant crumble with clotted cream made from your own blackcurrants. That key lesson is: what's good for us is good for nature, and what's good for nature is good for us. And nowhere is this more evident than in the bustling, flourishing, flowering, fruiting and altogether glorious native British hedge.

Notes

Introduction: A Hedge in Wiltshire

1. Isabella Tree, 'We Need to Bring Back the Wildwoods of Britain to Fight Climate Change', *Guardian*, 26 November 2018, https://www.theguardian.com/commentisfree/2018/nov/26/wildwoods-britain-climate-change-northern-forest.
2. RSPB, 'A Diverse Habitat', 12 October 2003, https://www.rspb.org.uk/helping-nature/what-we-do/influence-government-and-business/farming/farm-hedges/a-diverse-habitat.
3. Centre for Ecology and Hydrology, 'Rejuvenation of Hedgerows', National Environment Research Council, 10 June 2015, https://www.ceh.ac.uk/sites/default/files/documents/HedgerowRejuvenationResearchProject_SummaryLeaflet_June15.pdf.
4. Buglife, 'Ancient and Species-Rich Hedgerows', 27 March 2006, https://www.buglife.org.uk/resources/habitat-management/ancient-and-species-rich-hedgerows.
5. Dave Goulson, 'The Insect Apocalypse: "Our world will grind to a halt without them"', *Guardian*, 25 July 2021, https://www.theguardian.com/environment/2021/jul/25/the-insect-apocalypse-our-world-will-grind-to-a-halt-without-them.

Chapter One. Hedge-Priests, Hedge-Taverns and Hedge-Mumpers: A Brief History of the English Hedge

1. Richard and Nina Muir, *Hedgerows* (London: Michael Joseph, 1987), 3.
2. Must Farm, 'Dead Hedge', accessed September 2023, http://www.mustfarm.com/290/dead-hedge.
3. Muir, *Hedgerows*, 10.
4. Julius Caesar, *Gallic War* 2.17, trans. W.A. McDevitte and W.S. Bohn (New York. Harper and Brothers, 1869).
5. Campaign to Protect Rural England, 'England's Hedgerows: Don't Cut Them Out!', 30 August 2010, https://www.cpre.org.uk/resources/englands-hedgerows-dont-cut-them-out.
6. Muir, *Hedgerows*, 17–18.
7. Richard Muir, *Woods, Hedgerows and Leafy Lanes* (Stroud: Tempus, 2008), 17.
8. Knepp, 'Nightingales', 20 December 2022, https://knepp.co.uk/rewilding/wildlife-successes/nightingales.
9. Muir, *Hedgerows*, 30.
10. Ernest Pollard, Max D. Hooper and Norman W. Moore, *Hedges* (London: Collins, 1974), 77.

11. The Land is Ours, 'Robert Ket and the Norfolk Rising. 1549', 31 January 2001, https://tlio.org.uk/robert-ket-and-the-norfolk-rising-1549.

12. Muir, *Hedgerows*, 30.

13. Doug Campbell, 'Gerrard Winstanley, Excerpts From "The True Levellers' Standard Advanced" (1649)', Professor Campbell, 10 January 2020, http://www.professorcampbell.org/sources/winstanley.html.

14. John W. Dover, 'Introduction to Hedgerows and Field Margins' in *The Ecology of Hedgerows and Field Margins*, ed. John W. Dover (Abingdon: Routledge, 2019), 3.

15. David Streeter and Rosamond Richardson, *Discovering Hedgerows* (London: BBC Books, 1982), 13.

16. Streeter and Richardson, *Discovering Hedgerows*, 146–47.

17. Michael Allaby, *The Woodland Trust Book of British Woodlands* (Newton Abbot: David and Charles, 1986), 107.

18. Oliver Rackham, *The Illustrated History of the Countryside* (London: Weidenfeld and Nicolson, 1994), 13.

19. Dover, 'Introduction to Hedgerows', 2.

20. Reg Fry and David Lonsdale, eds., *Habitat Conservation for Insects: A Neglected Green Issue* (London: Amateur Entomologists' Society, 1991), 21:121.

21. Dover, 'Introduction to Hedgerows', 15.

22. George Monbiot, *Feral* (London: Penguin, 2014), 162.

23. Kieran Bell, 'Ancient Hedgerow Destroyed by Vearse Farm Developers', *Bridport and Lyme Regis News*, 1 December 2022, https://www.bridportnews.co.uk/news/23161459.ancient-hedgerow-destroyed-vearse-farm-developers.

24. John Baker, 'Locals Protest Over Ancient Hedgerows Being Destroyed', *Wiltshire Times*, 14 May 2023, https://www.wiltshiretimes.co.uk/news/23515467.locals-protest-ancient-hedgerows-destroyed, and, Annie Owen, 'Developer Responds After Concerns Over Cutting of Ancient Hedgerows to Make Way for Houses', *Daily Post*, 9 February 2022, https://www.dailypost.co.uk/news/north-wales-news/developer-responds-after-concerns-over-23025886.

25. Streeter and Richardson, *Discovering Hedgerows*, 145.

26. Sue Clifford and Angela King, *England in Particular* (London: Saltyard Books, 2006), 420–21.

27. Clifford and King, *England in Particular*, 420–21.

28. Rackham, *History of the Countryside*, 15.

Chapter Two. The Green and Brimming Bank: Hedges and Plants

1. John Wright, *A Natural History of the Hedgerow* (London: Profile Books, 2016), 126.

2. Ernest Pollard, quoted in Donald Grose, *The Flora of Wiltshire* (Devizes: Wiltshire Archaeological and Natural History Society, 1957), 73.

3. Richard Muir, *Woods, Hedgerows and Leafy Lanes* (Stroud: Tempus, 2008), Richard Muir, 143.

4. Peter Wohlleben, *The Hidden Life of Trees* (London: William Collins, 2017), vii.
5. Ernest Pollard, Max D. Hooper and Norman W. Moore, *Hedges* (London: Collins, 1974), 148.
6. The article is available to read free online here: Keith Alexander, Jill Butler and Ted Green, 'The Value of Different Tree and Shrub Species to Wildlife', The Arboricultural Association, updated 21 July 2021, https://trees.org.uk/News-Blog/Latest-News/The-value-of-tree-and-shrub-species-to-wildlife.
7. Pollard, Hooper and Moore, *Hedges*, 147.
8. Benedict Macdonald, *Cornerstones* (London: Bloomsbury Wildlife, 2022), 148.
9. Philip J. Wilson, 'Botanical Diversity in the Hedges and Field Margins of Lowland Britain' in *The Ecology of Hedgerows and Field Margins*, ed. John W. Dover (Abingdon: Routledge, 2019), 37.
10. Heather Tanner and Robin Tanner. *Wiltshire Village* (London: Garton, 1939), 119.
11. *Online Etymology Dictionary*, s.v. 'hag', updated 28 September 2017, https://www.etymonline.com/word/hag.
12. Oliver Rackham, *The Illustrated History of the Countryside* (London: Weidenfeld and Nicolson, 1994), 147.
13. Rebecca Oaks and Edward Mills, *Coppicing and Coppice Crafts* (Ramsbury: Crowood Press, 2010), 98.
14. Wilson, 'Botanical Diversity', 39.
15. Barnes Common, 'Pollarding', 26 January 2021, https://barnescommon.org.uk/conservation/habitat-management/techniques-practices/pollarding.
16. Barnes Common, 'Pollarding'.
17. Woodland Trust, 'Deadwood', 22 January 2019, https://www.woodlandtrust.org.uk/trees-woods-and-wildlife/habitats/deadwood.
18. The Wildlife Trusts, 'Noble Chafer', 28 May 2018, https://www.wildlifetrusts.org/wildlife-explorer/invertebrates/beetles/noble-chafer.
19. Rackham, *History of the Countryside*, 54.
20. Peter Marren, *Bugs Britannica*, ed. Richard Mabey (London: Chatto and Windus, 2010), 345–46.
21. Pollard, Hooper and Moore, *Hedges*, 149.
22. Pollard, Hooper and Moore, 153.
23. Richard and Nina Muir, *Hedgerows* (London: Michael Joseph, 1987), 155.
24. Alfred E. Hartemink, 'Chapter Two – The Definition of Soil Since the Early 1800s', *Advances in Agronomy* 137, (2016): 73–126, https://doi.org/10.1016/bs.agron.2015.12.001, quoted in Carmelo Dazzi and Giuseppe Lo Papa, 'A New Definition of Soil to Promote Soil Awareness, Sustainability, Security and Governance', *International Soil and Water Conservation Research* 10, no. 1 (March 2022): 99–108, https://doi.org/10.1016/j.iswcr.2021.07.001.
25. Rackham, *History of the Countryside*, 22.

Chapter Three. Nature's Hum: Hedges and Insects

1. John M. Holland, 'Contribution of Hedgerows to Biological Control' in *The Ecology of Hedgerows and Field Margins*, ed. John W. Dover (Abingdon: Routledge, 2019), 129.

2. Phoebe Weston, '"Reservoirs of Life": How Hedgerows Can Help the UK Reach Net Zero in 2050', *Guardian*, 2 February 2021, https://www.theguardian.com/environment/2021/feb/02/reservoirs-of-life-hedgerows-help-uk-net-zero-2050-aoe.

3. Jacques Baudry and Françoise Burel, 'Multi-Scale Control of Carabid Assemblages in Hedgerow Network Landscapes' in *The Ecology of Hedgerows and Field Margins*, ed. John W. Dover (Abingdon: Routledge, 2019), 149.

4. For a good overview, see: Jon Entine, 'Viewpoint: Is the Predicted "Silent Earth Insect Armageddon" the Inevitable Result of Using Farm Chemicals — Or is it Alarmist Activist Propaganda? Insect Scientists Challenge the Doomsayers', Genetic Literacy Project, 9 June 2023, https://geneticliteracyproject.org/2023/06/09/viewpoint-silent-earth-and-insect-armageddon-caused-by-farm-chemicals-or-activist-propaganda-insect-scientists-challenge-the-doomsayers.

5. Dave Goulson, *Silent Earth* (London: Jonathan Cape, 2021), 141.

6. Ernest Pollard, Max D. Hooper and Norman W. Moore, *Hedges* (London: Collins, 1974), 165.

7. Amanda E Martin, et al., 'Effects of Farmland Heterogeneity on Biodiversity Are Similar to—or Even Larger Than—the Effects of Farming Practices', *Agriculture, Ecosystems & Environment* 288 (February 2020): 106698, https://www.sciencedirect.com/science/article/pii/S0167880919303147.

8. Both cited in: Vicki Hird, *Rebugging the Planet* (London: Chelsea Green, 2021), 32.

9. I've Never Killed a Pipit, 'The Extinct Wildlife of The UK: 4 – The Blue Stag Beetle', 25 February 2013, https://iveneverkilledapipit.wordpress.com/2013/02/25/the-extinct-wildlife-of-the-uk-4-the-blue-stag-beetle.

10. I've Never Killed a Pipit, 'The Extinct Wildlife of the UK – 1: Spotted Sulphur', 4 February 2013, https://iveneverkilledapipit.wordpress.com/2013/02/04/the-extinct-wildlife-of-the-uk-1-spotted-sulphur.

11. Kathryn Smith, 'Case Study: British Dung beetles – Here to Help', Farm Wildlife, 15 April 2019, https://farmwildlife.info/2019/04/15/british-dung-beetles-here-to-help.

12. Patrick Barkham, 'Ants Run Secret Farms on English Oak Trees, Photographer Discovers', *Guardian*, 24 January 2020, https://www.theguardian.com/environment/2020/jan/24/ants-run-secret-farms-on-english-oak-trees-photographer-discovers.

13. Society for Promoting Christian Knowledge, *The Home Friend, A Weekly Miscellany of Amusement and Instruction* (London: Society for Promoting Christian Knowledge, 1855), 3:300.

14. Robert Mudie, *The Feathered Tribes of the British Islands* (London: Whittaker and Co., 1834), 1:200.

15. Pollard, Hooper and Moore, *Hedges*, 189.
16. Toby Doyle, et al., 'Pollination by Hoverflies in the Anthropocene', *Proceedings of the Royal Society B: Biological Sciences 287*, no. 1927 (May 2020): 20200508, http://dx.doi.org/10.1098/rspb.2020.0508.
17. Oliver Milman, *The Insect Crisis* (London: Atlantic Books, 2022) 6.
18. Doyle, et al., 'Pollination by Hoverflies', 20200508.
19. Doyle, et al., 20200508.
20. Claire Kremen, Matthias Albrecht and Lauren Ponisio, 'Restoring Pollinator Communities and Pollination Services in Hedgerows in Intensively Managed Agricultural Landscapes' in *The Ecology of Hedgerows and Field Margins*, ed. John W. Dover (Abingdon: Routledge, 2019), 163.
21. Cristina Botías, et al., 'Impact of Pesticide Use on the Flora and Fauna of Field Margins and Hedgerows' in *The Ecology of Hedgerows and Field Margins*, ed. John W. Dover (Abingdon: Routledge, 2019), 101.
22. Holland, 'Contribution of Hedgerows', 125.
23. Ellen A.R. Welti, et al., 'Nutrient Dilution and Climate Cycles Underlie Declines in a Dominant Insect Herbivore', *Proceedings of the National Academy of Sciences* 117, no. 13 (March 2020): 7,271–75, https://doi.org/10.1073/pnas.1920012117.
24. Oliver Milman, 'Bees Can Play Soccer – 10 Little-Known Facts about Insects', *Guardian*, 1 March 2022, https://www.theguardian.com/environment/2022/mar/01/insect-crisis-book-climate-bees-moths-roaches-flies-facts.

Chapter Four. A Bustle in Your Hedgerow: Birds, Hedgehogs and Tree Frogs

1. Richard Muir, *Woods, Hedgerows and Leafy Lanes* (Stroud: Tempus, 2008), 204.
2. Benedict Macdonald, *Rebirding* (London: Pelagic Publishing, 2019), 71.
3. Macdonald, *Rebirding*, 72.
4. David Streeter and Rosamond Richardson, *Discovering Hedgerows* (London: BBC Books, 1982), 127.
5. John Wright, *A Natural History of the Hedgerow* (London: Profile Books, 2016), 271.
6. Patrick Vallance, 'We've Overexploited the Planet, Now We Need to Change if We're to Survive', Government Office for Science, 8 July 2022, https://www.gov.uk/government/speeches/weve-overexploited-the-planet-now-we-need-to-change-if-were-to-survive.
7. Joint Nature Conservation Committee, 'UKBI – C5. Birds of the Wider Countryside and at Sea', 14 November 2023, https://jncc.gov.uk/our-work/ukbi-c5-birds-of-the-wider-countryside-and-at-sea.
8. Department for Environment Food & Rural Affairs, 'Wild Bird Populations in the UK, 1970 to 2022', National Statistics, updated 7 November 2023, https://www.gov.uk/government/statistics/wild-bird-populations-in-the-uk/wild-bird-populations-in-the-uk-1970-to-2021.

9. Mark Cocker and Richard Mabey, *Birds Britannica* (London: Chatto and Windus, 2005), 304.

10. Macdonald, *Rebirding*, 69.

11. Cocker and Mabey, *Birds Britannica*, 185.

12. Tom Moorhouse, *Ghosts in the Hedgerow* (London: Doubleday, 2023), 93.

13. Royal Society of Biology, 'Hedgehog Wins Favourite UK Mammal Poll', 29 November 2016, https://www.rsb.org.uk/news/hedgehog-wins-favourite-uk-mammal-poll.

14. George C. Kotzageorgis and Christopher F. Mason, 'Small mammal populations in relation to hedgerow structure in an arable landscape', *Journal of Zoology* 242, no. 3 (July 1997): 425–34, https://doi.org/10.1111/j.1469-7998.1997.tb03846.x.

15. Thomas Browne, *Pseudodoxia Epidemica* (London: T.H. for Edward Dod, 1646), 138.

16. Harry Cockburn, 'Rewilding: Can Britain's Long Lost Tree Frogs Bounce Back?', *Independent*, 3 February 2021, https://www.independent.co.uk/news/uk/home-news/tree-frogs-rewilding-biodiversity-beavers-b1797072.html.

Chapter Five. Hedges: The Living Food Bank That Keeps On Giving

1. John Lewis-Stempel, *The Wild Life* (London: Doubleday, 2009), 197.

2. David Streeter and Rosamond Richardson, *Discovering Hedgerows* (London: BBC Books, 1982), 51.

3. Richard Mabey, *Flora Britannica* (London: Chatto and Windus, 1996), 152–53.

4. Anna Magiera, et al., 'Polyphenol-Enriched Extracts of *Prunus spinosa* Fruits: Anti-Inflammatory and Antioxidant Effects in Human Immune Cells Ex Vivo in Relation to Phytochemical Profile', *Molecules* 27, no. 5 (March 2022): 1691, https://doi.org/10.3390/molecules27051691.

5. *Jackson's Oxford Journal* (Oxford: William Jackson, 1778), quoted in John Wright, *A Natural History of the Hedgerow* (London: Profile Books, 2016), 6.

6. Andy Hamilton, *Booze for Free* (London: Transworld, 2011), 154.

7. Janice Pattie, 'How to Make Hedgerow Vodka', Farmersgirl Kitchen, 14 October 2017, https://www.farmersgirlkitchen.co.uk/how-to-make-hedgerow-vodka.

8. Alfred Williams, *Life in a Wiltshire Village* (Gloucester: Alan Sutton, 1981), 162.

9. Mabey, *Flora Britannica*, 388.

Chapter Six. Hedges as Carbon Banks

1. Sophie Drexler, Andreas Gensior and Axel Don, 'Carbon Sequestration in Hedgerow Biomass and Soil in the Temperate Climate Zone', *Regional Environmental Change* 21, no. 74 (July 2021), https://link.springer.com/article/10.1007/s10113-021-01798-8.

2. Chelsea Harvey and E&E News, 'The World's Winds Are Speeding Up', *Scientific American*, 19 November 2019, https://www.scientificamerican.com/article/the-worlds-winds-are-speeding-up.

3. Richard Muir, *Woods, Hedgerows and Leafy Lanes* (Stroud: Tempus, 2008), 226.

4. Hedgelink, 'Importance of Hedgerows', 22 Sept 2008, https://hedgelink.org.uk/guidance/importance-of-hedgerows.
5. John W. Dover, 'Introduction to Hedgerows and Field Margins' in *The Ecology of Hedgerows and Field Margins*, ed. John W. Dover (Abingdon: Routledge, 2019), 17–18.
6. Lyndsey Graham, et al., 'Remote Sensing Applications for Hedgerows' in *The Ecology of Hedgerows and Field Margins*, ed. John W. Dover (Abingdon: Routledge, 2019), 83.
7. Ross Dickinson, 'A Coppiced Hedge: Converting a Flailed Hedge into an Economic Crop of Firewood', Farm Woodland Forum, 13 June 2018, https://www.agroforestry.ac.uk/resources/case-studies/coppiced-hedgerows-racedown-farm.
8. Bill McGuire, *Hothouse Earth* (London: Icon Books, 2022), 153.

Chapter Seven. Pleachers, Bodgers, Brummocks and Flailers

1. Benedict Macdonald and Nicholas Gates, *Orchard* (London: William Collins, 2020), 75.
2. Macdonald and Gates, *Orchard*, 75.
3. Isabella Tree and Charlie Burrell, *The Book of Wilding* (London: Bloomsbury, 2023), 94.
4. Tree and Burrell, *Wilding*, 155–56.
5. Benedict Macdonald, *Cornerstones* (London: Bloomsbury Wildlife, 2022), 35.

Further Reading

Allaby, Michael. *The Woodland Trust Book of British Woodlands*. Newton Abbot: David and Charles, 1986.

Clifford, Sue and Angela King. *England in Particular*. London: Saltyard Books, 2006.

Cocker, Mark and Richard Mabey. *Birds Britannica*. London: Chatto and Windus, 2020.

Dover, John W., ed. *The Ecology of Hedgerows and Field Margins*. Abingdon: Routledge, 2019.

Goulson, Dave. *Silent Earth*. London: Jonathan Cape, 2021.

Hamilton, Andy. *Booze for Free*. London: Transworld, 2011.

Lewis-Stempel, John. *The Wild Life*. London: Doubleday, 2009.

Mabey, Richard. *Flora Britannica*. London: Chatto and Windus, 1996.

Mabey, Richard. *Food for Free*. London: William Collins, 2022.

Macdonald, Benedict. *Cornerstones*. London: Bloomsbury Wildlife, 2022.

Macdonald, Benedict and Nicholas Gates. *Orchard*. London: William Collins, 2020.

Macdonald, Benedict. *Rebirding*. London: Pelagic Publishing, 2019.

McGuire, Bill. *Hothouse Earth*. London: Icon Books, 2022.

Monbiot, George. *Feral.* London: Penguin, 2014.

Moorhouse, Tom. *Ghosts in the Hedgerow.* London: Doubleday, 2023.

Muir, Richard and Nina Muir. *Hedgerows.* London: Michael Joseph, 1987.

Muir, Richard. *Woods, Hedgerows and Leafy Lanes.* Stroud: Tempus, 2008.

Pollard, Ernest, Max D. Hooper and Norman W. Moore. *Hedges.* London: Collins, 1974.

Rackham, Oliver. *The Illustrated History of the Countryside.* London: Weidenfeld and Nicolson, 1994.

Streeter, David and Rosamond Richardson. *Discovering Hedgerows.* London: BBC Books, 1982.

Tanner, Heather and Robin Tanner. *Wiltshire Village.* London: Garton, 1939.

Thomson, Jonathan. How to Rewild. Bridport: Dovecote Press, 2021.

Tree, Isabella and Charlie Burrell. *The Book of Wilding.* London: Bloomsbury, 2023.

Williams, Alfred. *Life in a Wiltshire Village.* Gloucester: Alan Sutton, 1981.

Wohlleben, Peter. *The Hidden Life of Trees.* London: William Collins, 2017.

Wright, John. *The Forager's Calendar.* London: Profile Books, 2019.

Wright, John. *A Natural History of the Hedgerow.* London: Profile Books, 2016.

Index

Acknowledgements

I owe a huge debt of thanks to Harry James, the ecologist who first made a study of the hedge at Underhill, which is the centrepiece of this book, and established what a tremendous difference good hedgerow management can make to bio-diversity. Thanks also to the invaluable contribution of Mark Arbuckle, who carried out a very detailed invertebrate survey of the hedge, and Jenny Bennett, who carried out the full botanical survey. All of this hard science has, I hope, helped to give the book a real backbone. I would also like to thank my friend the wonderful landscape artist Richard Hoare for allowing me to use two of his brilliant photos of the Knepp Estate, which perfectly illustrate the value of thickets and suggest the sort of ancient wood-pasture landscape that would have been so favourable to them.

Thank you also to Muna Reyal at Chelsea Green for her editing and her patience, to Susan Pegg for her brilliant copy-editing, and to the whole team there. It's been a great support.

And, last but not least, to my wife, Iona, who has listened to me chattering on about all things hedge-related for months on end, and yet still appears to be genuinely interested. A constant encouragement!

Biographies

About Christopher Hart

Christopher has written ten novels to date, including the literary works *Lost Children* and *The Harvest*. Published under the pen name of William Napier, his historical fiction includes *Julia*, the best-selling *Attila* trilogy and *The Last Crusaders* trilogy. His work has been praised in both *The Times Literary Supplement* and the *Sunday Sport*. He has also published numerous short stories, essays and reviews. Chris has been a freelance journalist since the 1990s, writing for *The Sunday Times, Daily Mail, Daily Telegraph* and others. He lives in Wiltshire, where he is rewilding seven acres and a hedge.

About Jonathan Thomson

Jonathan has run a twenty-five-acre rewilding project called Underhill Wood Nature Reserve in Wiltshire since 2014 for flora, fauna and the education of young people. There, he has restored a hedge to its former glory through the practice of conservation hedge-laying. Jonathan teaches the John Muir Conservation Award and runs rewilding workshops. He grew up on a small dairy farm in rural New Zealand, and the natural world has always played an important part in his life. Since living in the UK, his awareness of the need to protect, conserve and restore our squeezed and threatened native fauna and flora has grown more urgent. www.underhillwoodnaturereserve.com.